Torsion of Thin Walled Structures

Krishnaiyengar Rajagopalan

Torsion of Thin Walled Structures

 Springer

Krishnaiyengar Rajagopalan
Department of Ocean Engineering
Indian Institute of Technology Madras
Chennai, Tamil Nadu, India

ISBN 978-981-16-7460-0   ISBN 978-981-16-7458-7   (eBook)
https://doi.org/10.1007/978-981-16-7458-7

This Springer imprint is published by the registered company Springer Nature Singapore Pte Ltd.
The registered company address is: 152 Beach Road, #21-01/04 Gateway East, Singapore 189721,
Singapore

Dedicated to

श्री गुरुवायूरप्पन् स्वामी

Shri Guruvayoorappan Swamy

Srimad Bhagavad Gita Chapter 18, Verse 46

यतः प्रवृत्तिर्भूतानाम् येन सर्वमिदं ततम्
स्वकर्मणा तमभ्यर्च्य सिद्धिं विंदति मानवः

*yataḥ pravṛittir bhūtānāṁ yena sarvam idaṁ
tatam
sva-karmaṇā tam abhyarchya siddhiṁ
vindati mānavaḥ*

*Man attains the highest perfection by
worshipping him with his own natural duties,
from whom all this universe springs forth and
by whom all this universe is pervaded*
(Translation source: https://www.holy-bha
gavad-gita.org/)

Preface

Thin-walled structures in which torsion is dominant occur in ship and aircraft structures, shear walls and bridges, ladder frames of commercial automobile vehicles, satellite support systems, and the like. These structures warp under torsion, and such warping complicates the analysis. The elucidation of warping properties is basic to the understanding of the subject, but this topic is not well addressed in the existing books.

Considerable attention is therefore devoted in this book to make this somewhat esoteric problem exoteric, and the author considers this as the forte of the book.

Review questions and problems with full answers further embellish the portrayal of the subject. Classical and finite element techniques are used in the analysis, and wherever possible, the same example is solved by both methods so that the reader can come to grips with the subject.

The subject is a classical as well as a modern one. The references given at the end of the chapters were freely consulted, gleaned, and circulated in classroom situations. The author wishes to thank the writers of these references. Nevertheless, efforts have been made to present the classical concepts in lucid terms so as to dovetail them with modern finite element concepts for emphasizing the importance of torsion in thin-walled structures.

Chennai, India

Krishnaiyengar Rajagopalan
01558@retiree.iitm.ac.in

Contents

About the Author

Dr. Krishnaiyengar Rajagopalan obtained his bachelor's and master's in Civil Engineering and Structural Engineering, respectively, from the University of Madras, India. He pursued his Ph.D. from the Indian Institute of Technology (IIT) Madras for his research on the finite element buckling analysis of thin walled shell structures. Dr. Rajagopalan has worked as a scientist with CSIR, India and as a professor with IIT Madras. He worked extensively in the areas of thin walled structures mainly in the classical and finite element analysis of ship structures, power plant structures, storage structures, calandria (nuclear) vessels, boiler supporting structures, submarine structures, steel and concrete building structures and of similar ilk.

Chapter 1
Torsion of Thin-Walled Structures

1.1 Introduction

Thin-walled structures are extensively used in several branches of engineering. Basically, these are structures which consist of thin straight or curved plates joined at their edges. The thickness of the plates is small compared to the cross-sectional dimensions and span. Such structures are commonly used in ships, aircrafts, automobiles, submarines, offshore tubular structures, cores of tall buildings, crane and bridge girders, machine tool structures, satellite support systems, and the like. Although the material is usually steel, thin-walled structures of reinforced and prestressed concrete, aluminum, and fiber-reinforced plastic (FRP) also exist. The primary advantage of these structures is their lightweight. Some commonly occurring thin-walled structures are shown in Fig. 1.1.

The following features differentiate a thin-walled structure from the thick-walled counterparts.

a. Shear stresses and strains are larger in thin-walled structures.
b. In-plane stresses can induce local buckling in thin-walled structures.
c. St. Venant's principle is not valid, i.e., if self-equilibrating forces are applied over a region, the stresses do not attenuate rapidly along the length away from the region.

Thin-walled structures when twisted result in the development of the warping of the cross section. The term warping denotes out-of-plane distortion (deplanation) of the cross section in the direction of the axis of the structure. When warping is inhibited by restraints at internal points or at the boundaries of the structure, direct (normal) stresses in the axial direction and shear stresses in the cross section are introduced. These stresses are called warping stresses which are additional to those stresses caused by axial and bending loads. The decoupling of the axial, bending and warping effects enables the evaluation of the normal stress at any point in the cross section as [1]

© The Author(s), under exclusive license to Springer Nature Singapore Pte Ltd. 2022   1
K. Rajagopalan, *Torsion of Thin Walled Structures*,
https://doi.org/10.1007/978-981-16-7458-7_1

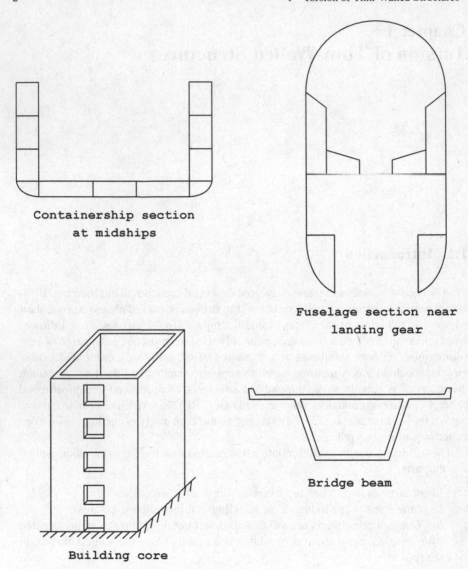

Fig. 1.1   Certain slim-walled structures

$$\sigma = \frac{P}{A_S} + \frac{M_X Y}{I_X} + \frac{M_Y X}{I_Y} + \frac{M_\Omega \omega}{I_\Omega} \qquad (1.1)$$

where X and Y are the principal planes, P is the axial load, and M_X, M_Y are the bending moments. The principal moments of inertia are I_X and I_Y. The fourth term in the foregoing equation gives the contribution to the normal stress due to the restraint of warping which results in the development of a bimoment M_Ω. The quantities

ω and I_Ω are cross-sectional properties (similar to sectional area of the material A_S and the moments of inertia I). They are called unit warping (dimensions, m^2) and warping moment of inertia (dimensions, m^6), respectively.

In a similar way, the contribution to shear stress by warping can be given by

$$\tau_W = \frac{T_\Omega S_\Omega}{t I_\Omega} \tag{1.2}$$

which is similar to the familiar equation for transverse shear stress

$$\tau = \frac{V Q}{t I} \tag{1.3}$$

In Eq. (1.2), the warping torsional moment (derivative of bimoment) resulting due to restraint of warping is denoted by T_Ω and S_Ω is the sectorial shear function (a cross-sectional property). It may be noted that the normal and shear stresses, σ_W and τ_W, due to warping arise wherever warping is restrained. These stresses will be zero at all places in the structure where the cross section is free to warp. The torsional analysis of thin-walled structures under uniform and non-uniform (warping) torsion is dealt with in detail in this book. Because of the specialized nature of the subject, much fruitful research has been done in the area over the years. Yet the problem is a practical one as there are many classes of structures wherein torsion is the primary loading such as the cross-frames in the chassis of automobile trucks, containership hulls, fuselages of aircrafts, shear cores in tall buildings, azimuth tubes supporting radio telescopes, spine beams of bridges, and the like.

1.2 Picture of Sectional Stresses

The distribution of shear stresses due to bending in principal planes can be computed using the formula VQ/It and has been covered in text books on mechanics of materials. A summary of the problem is presented in the appendix. The shear stress in a thin-walled section at any point is uniformly distributed in the thickness direction as shown in Fig. 1.2. The magnitude along the contour varies linearly in the flange and parabolically in the web as shown in the figure. The shear stress arises from equilibrium considerations as shown. If we assume bending involving compressive stresses in the top flange and tensile stresses in the bottom flange, the shear stress directions along the contour are as shown in Fig. 1.2. Because of the fact that the shear stress is distributed uniformly in the thickness direction, we can write $q = \tau t$ where τ is the shear stress (units N/m^2) and t is the thickness. The quantity q is called the shear flow and has unit of N/m. The flow of the shear from the top free edge to the bottom free edge is likened to the flow of a river. The shear flow in other types of thin-walled open sections can be visualized easily. A few examples are shown in Fig. 1.3. In each case, the bending is in the vertical principal plane.

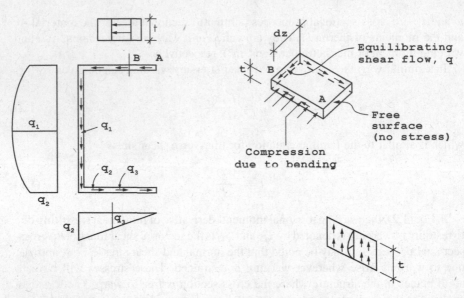

Fig. 1.2 Distribution of shear flow and Neglected shear stress in flange due to transverse shear

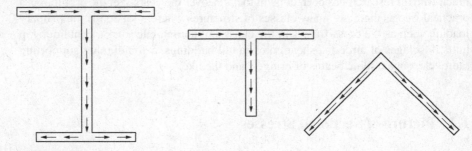

Fig. 1.3 Shear flows due to vertical transverse shear in open sections (symmetrical bending)

It must be noted that there is also a small vertical shear stress in the flange which is uniformly distributed along the contour of the flange. This stress arises from the formula VQ/Ib and is distributed parabolically in the thickness direction as shown in Fig. 1.2. This stress being very small is usually neglected.

The shear flows due to transverse shear in closed thin-walled cross sections can be visualized in a similar manner. A few examples are shown in Fig. 1.4. In unsymmetric sections, it is difficult to know in advance the point of zero shear, and hence, the shear flow cannot be determined from simple equilibrium considerations only. Compatibility conditions must be enforced to determine the shear flows. The procedure is given in the appendix.

The concept of the shear center is illustrated in Fig. 1.5. Each of the forces shown can be replaced by an equipollent force and moment at the point S. When the resultant

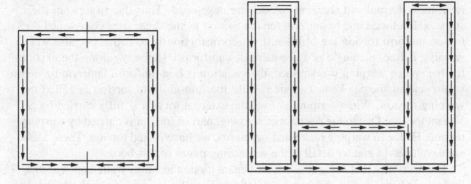

Fig. 1.4 Shear flows due to vertical transverse shear in closed sections (Symmetrical bending)

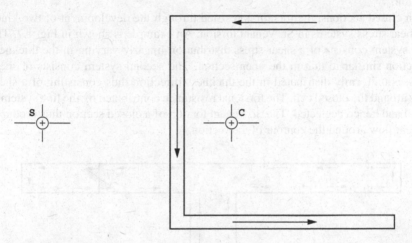

Fig. 1.5 Concept of shear center

yields only a vertical force and no moment (i.e., no torsion on the section), we achieve the condition of pure bending in the vertical plane (i.e., without any torsion). This position in the plane of the cross section is called the shear center. It is the point through which external loads must be applied to get torsion-less bending. (In the beam with the channel section, a horizontal bracket must be attached up to the point S, and the vertical loads must be applied at S. If loads are applied elsewhere, combined bending and torsion would result). The shear center is always located on the axis of symmetry if the section has any. The centroid is also located on this axis. For sections with two axes of symmetry, the shear center, S, coincides with the centroid C.

A picture of the shear stresses that arises in a thin-walled section due to transverse shear (i.e., bending) is given in the foregoing. Shear stresses can also result from pure torsion. Except circular or regular polygonal thin-walled tubes (Neuber tubes), most of the sections warp when torsion is applied. When the warping displacements are restrained, normal and shearing stresses are developed. Thus due to torsion, shear stresses developed due to uniform torsion as well as the shear stress developed due to non-uniform torsion are of interest. Uniform torsion is so-called because when warping is free, the angle of twist increases uniformly (linearly) along the axis of the bar. When warping is restrained, the variation is non-uniform. Uniform torsion is also called the St. Venant torsion while the non-uniform torsion is called the warping torsion. When warping is free, the external torsion is fully carried by St. Venant torsion. Due to the restraint of warping, part of torsion is carried by warping torsion. Hence in warping restrained structures, we have mixed torsion. These ideas are developed in greater detail in the succeeding pages of this book.

The distribution of shear stress in St. Venant torsion in a thin-walled open section is shown in Fig. 1.6. The shear stress effectively has a linear variation in the thickness direction. The flow of shear appears in the form of several parallel rivers within the section.

In closed sections, the torsion is resisted through the development of two kinds of shear stress systems in St. Venant torsion. An example is shown in Fig. 1.7. The first system consists of a shear stress distribution linearly varying in the thickness direction similar to that in the open section. The second system consists of shear stresses uniformly distributed in the thickness direction thus consisting of a shear flow around the closed cell. The torsional resistance contributed by the first system is small and hence neglected. The St. Venant torsion of a closed section thus produces a shear flow around the contour of the section.

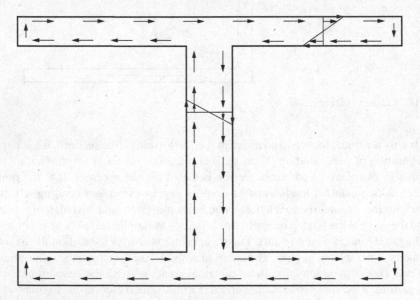

Fig. 1.6 St. Venant torsion in a thin-walled open section

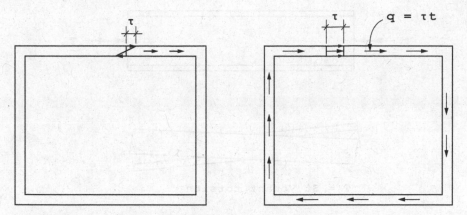

Fig. 1.7 Two shear stress systems in St. Venant torsion in a closed sections

Contours of open and closed sections warp when torsion is applied to them. (Exceptions are open sections which originate from a point such as angle, tee, cruciform, and star and closed regular polygonal single-cell tubes). Warping represents displacements normal to the plane of the cross section. At a support where these displacements are restrained, warping normal and shear stresses develop, and the entire applied torsion is resisted as warping torsion. At the fully fixed section, St. Venant system of shear stresses will be absent. As shown in Fig. 1.8, when warping displacements are free to occur as at the free end of the bar, the entire torsion is resisted as St. Venant torsion and warping normal and shear stress systems will be absent here. For sections in between, the applied torque is resisted partly by St. Venant torsion and partly by warping torsion. At these sections, both the St. Venant and warping stress systems will be present.

Warping consists of out-of-plane displacements and arises due to the bending of the flanges in opposite directions as shown in Fig. 1.9. The section is subjected to two bending moments M which system is self-equilibrating. The product of the bending moment and the distance between the flanges ($M_\Omega = Mh$) is called the bimoment. The bimoment M_Ω is responsible for the development of the normal stress due to warping as shown in Eq. (1.1) and as can be visualized schematically in Fig. 1.9. Since the flange is under bending, a shear force must arise out of it. The shear stress due to these shear forces is uniformly distributed in the thickness direction and varies parabolically along the flange. These are the warping shear stresses. The equal and opposite shear force system constitutes a torsional moment which resists part of the applied torsion. This is the warping torsional moment T_Ω and warping shear stress developed due to T_Ω as shown in Eq. (1.2). The warping torsional moment depends on the shear force in the flange which is the derivative of M as shown in textbooks on simple beam theory. Since

$$T_\Omega = Vh = \frac{dM}{dz}h = \frac{d}{dz}(Mh) = \frac{dM_\Omega}{dz}$$

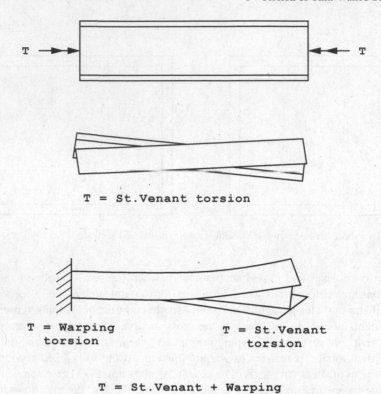

Fig. 1.8 Warping of a cantilever

Hence, the warping torsional moment is simply the first derivative of the bimoment.

1.3 Decoupling of Bending and Warping

The decoupling of axial, bending, and warping effects can be shown with the aid of Fig. 1.10. The load of magnitude $4P$ acting at the end of the flange can be replaced by an axial load of $4P$ at the centroid, a bending moment $2Ph$ about the y-axis, a bending moment $2Pb$ about the x-axis, and a bimoment Pbh. The axial and bending effects can be treated by the usual theories.

The warping loads form a self-equilibrating system. However, each flange bends under a moment Pb, and hence, axial stresses and shear stresses are developed. As shown in Fig. 1.11, the bending of the flanges also results in the twisting of the section.

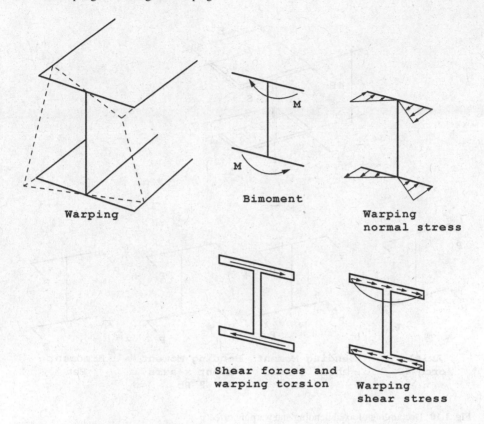

Fig. 1.9 Warping of flanged beam

The decoupling of bending and torsion is achieved by the application of transverse (bending) loads through the shear center. Thus, the decoupled system would involve the following.

i. Axial force at the centroid
ii. Bending moments about principal axes applied at the centroid
iii. Shear forces at the shear center
iv. Torsional moment applied at the shear center
v. Bimoment which does not have a location.

The theme of the book centers around (iv) and (v) in the foregoing. However, decoupling is possible only in straight beams and linear analyses. Thus, it is usual to consider all the effects together and take advantage of decoupling in particular cases.

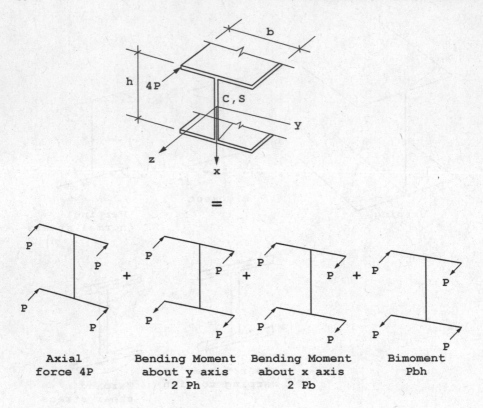

Fig. 1.10 Decoupling of axial bending and warping effects

1.4 Historical Overview

Historically, bending and torsion were investigated separately. The Navier–Bernoulli–Euler assumption of plane section remaining plane (i.e., warping or deplanation does not occur) gave the engineering theory of bending of thin beams. The shear stress in such beams is given by Jourawski (Eq. 1.3).

Torsion was investigated by St. Venant based on the view that the cross section is free to warp. The membrane analogy given by Prandtl and the work of Batho and Bredt extended the theory to closed (monocoque) tubes. Maillart presented the concept of shear center for the decoupling of bending and torsion.

Warping torsion was investigated by Timoshenko, Vlasov, Wagner, Bleich, and others. The theory for bars of open section developed by Vlasov was extended to torsion of closed sections by Von Karman and Christensen, Sadlecek, and others. More refined theories for closed sections were developed by Benscoter and by Kollbrunner and Hajdin.

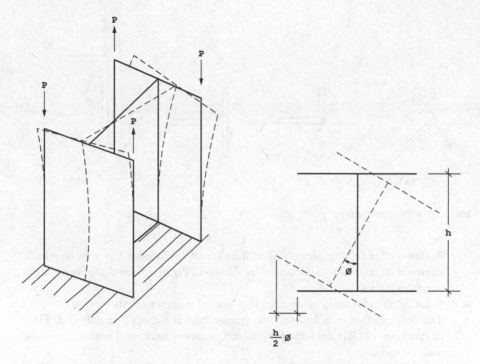

Fig. 1.11 Twisting of the section due to bending of the flanges

Recent developments have concentrated on other fundamental problems such as plastic torsion, torsional buckling, curved bars, and the like. Several contributions have been made in the last two decades on a variety of applications such as bridge beams and building cores, ship hull structures, mechanical engineering structures such as machine tool frames and turbine blades, aircrafts structures, automobile structures, space structures, and other thin-walled structures of similar ilk. It is the aim of this book to present the methodologies for the computation of warping torsion in a systematic manner followed by their applications to thin-walled structures encountered in various branches of engineering.

1.5 Review Problems

1. For each of the sections shown in Fig. 1.12, state whether the section deflects horizontally and/or twists under the vertical load indicated. The section shown is the end section of a cantilever.
2. Explain the physical meaning of shear center.

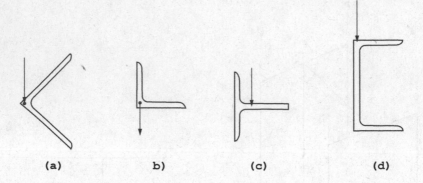

(a)   b)   (c)   (d)

Fig. 1.12 Review problem 1

3.   A thin-walled tubular section of uniform wall thickness t has a hole which subtends an angle 2α at the center as shown in Fig. 1.13. Find the location of the shear center.
4.   What are the two ways by which a thin-walled member resists torsion?
5.   The cross section of a thin-walled circular tube is slit open lengthwise. Find as functions of (R/t) the factors by which the shear stress and rate of torsional rotation increase.
6.   A thin-walled closed tube of length L has a constant wall thickness t. The radius however increases gently from r_o to r_L over the length L. Find an expression for the relative angle of torsional rotation between the ends.

Fig. 1.13 Review problem 3

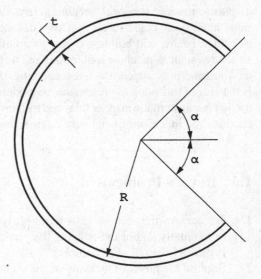

1.6 Answers to Review Problems

1. (a) deflects down. No horizontal deflection. No twist
    (b) Deflects down and to the left without twisting
    (c) Deflects down and twists
    (d) Deflects down and twists.
2. When lateral loads are applied over the length of the beam, they cause not
    only bending of the beam but also twisting. The resultant of the flexural shear
    stress (VQ/It) at any point on the cross section, such as the centroid, is not only a
    shear force but also a torsional moment. The shear stress system would however
    reduce only to a resultant shear force at a particular point on the cross section.
    This point is called the shear center. When lateral loads are applied to the beam
    through this point, only bending of the beam will result, and there will thus be
    no twisting.
3. The center of gravity of the segmental section is located at a distance g as shown
    in Fig. 1.14 where $g = \frac{R \sin \alpha}{\pi - \alpha}$.

    The moment of inertia of the section

    $$I = 2 \int_{\alpha}^{\pi} [Rtd\theta]R^2 \sin^2\theta$$

    $$= R^3 t \left[\pi - \alpha + \frac{\sin 2\alpha}{2} \right]$$

    The shear stress at section θ is

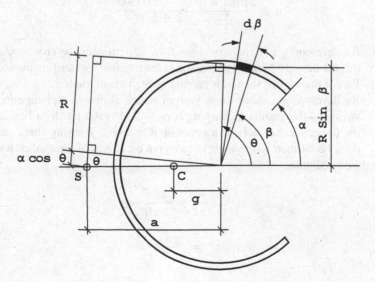

Fig. 1.14 Answer to review problem 3

$$\tau = \frac{VQ}{It}$$

where

$$Q = \int_{\alpha}^{\theta} [Rtd\beta][R\sin\beta]$$

$$= R^2 t[\cos\alpha - \cos\theta]$$

Thus,

$$\tau = \frac{V}{Rt} \frac{\cos\alpha - \cos\theta}{\left[\pi - \alpha + \frac{\sin 2\alpha}{2}\right]}$$

The resultant moment of the shear stresses about the shear center located at distance a from the center of the section must be zero. Hence,

$$2\int_{0}^{\pi} \frac{V}{Rt} \frac{\cos\alpha - \cos\theta}{\left[\pi - \alpha + \frac{\sin 2\alpha}{2}\right]}[Rtd\theta][R + a\cos\theta] = 0$$

i.e. $\dfrac{2V}{\left[\pi - \alpha + \frac{\sin 2\alpha}{2}\right]} \left\{ R[\sin\alpha + (\pi - \alpha)\cos\alpha] - a\left[\dfrac{\pi - \alpha}{2} + \dfrac{\sin 2\alpha}{4}\right]\right\} = 0$

which gives

$$a = \frac{2R[\sin\alpha + (\pi - \alpha)\cos\alpha]}{\left[\pi - \alpha + \frac{\sin 2\alpha}{2}\right]}$$

4. (a) By developing a circulatory shear flow system over the cross-section—this results in St. Venant torsion and is responsible for twisting the section. Part of the applied torsion is carried as St. Venant torsion.

    (b) By developing a shear stress system which is due to a change in axial stresses—this results in warping torsion and is the result of bending of the flanges of the section in opposite directions. Warping shear stresses do not contribute to the angle of twist but carry part of the applied torsion.

5. For the closed tube,

$$\phi'_C = \frac{T}{GJ} = \frac{T}{2\pi R^3 tG}$$

$$\tau^C = \frac{T}{2\pi R^2 t}$$

For the open tube

$$J = \frac{1}{3}(2\pi R)t^3 = \frac{2\pi Rt^3}{3}$$

$$\phi_0' = \frac{T}{GJ} = \frac{3T}{2\pi Rt^3 G}$$

$$\tau^\circ = \frac{Tt}{J} = \frac{3T}{2\pi Rt^2}$$

$$\frac{\tau^C}{\tau^\circ} = \frac{1}{3}\left(\frac{t}{R}\right)$$

$$\frac{\phi_C'}{\phi_0'} = \frac{1}{3}\left(\frac{t}{R}\right)^2$$

$$\text{i.e.} \quad \tau^\circ = 3\left(\frac{R}{t}\right)\tau^C$$

$$\phi_0' = 3\left(\frac{R}{t}\right)^2 \phi_C'$$

6.   We have the radius at any section

$$r = \xi r_L + (1 - \xi)r_0$$

where $\xi = 0$ and $\xi = 1$ denote the ends of the tube in which

$$\xi = \frac{z}{L}.$$

$$\phi' = \frac{d\phi}{dz} = \frac{T}{GJ} = \frac{T}{2\pi Gt\{\xi r_L + (1 - \xi)r_0\}^3}$$

Thus,

$$\phi = \int_0^1 \frac{T L d\xi}{2\pi Gt\{\xi r_L + (1 - \xi)r_0\}^3}$$

$$= \frac{TL}{2\pi r_0^3 Gt} \int_0^1 \frac{d\xi}{\left\{1 - \xi + \left(\frac{r_L}{r_0}\right)\xi\right\}^3}$$

Let

$$\rho = 1 - \xi + \frac{r_L}{r_0}\xi$$

$$\mathrm{d}\rho = -\left(1 - \frac{r_L}{r_0}\right)\mathrm{d}\xi$$

$$\begin{aligned}
\phi &= \frac{TL}{2\pi r_0^3 Gt} \int\limits_1^{r_L/r_0} \frac{-\mathrm{d}\rho}{\left\{1 - \frac{r_L}{r_0}\right\}\rho^3} \\
&= \frac{TL}{2\pi r_0^3 Gt} \frac{r_0(r_0 + r_L)}{2r_L^2} \\
&= \frac{TL}{4\pi Gt} \frac{r_0 + r_L}{r_0^2 r_L^2}
\end{aligned}$$

Reference

1. Hartog, D.: Advanced Strength of Materials. McGraw Hill, New York (1952)

Chapter 2
St. Venant Torsion

2.1 Introduction

A prismatic beam with a general thin-walled cross section that is allowed to warp freely when a torsional moment is applied to it is in a state of uniform torsion. This phenomenon is referred to as St. Venant torsion. As stated in the previous chapter, the prevention of warping introduces a state of non-uniform torsion in which a part of the total torsion is resisted by the warping torsion. The remainder is still carried by St. Venant torsion. It is the purpose of this chapter to discuss the St. Venant torsion in the case of cross sections with open profiles and multi-cellular profiles. Expressions for the shear stress distribution and torsional stiffness are presented with suitable examples which illustrate their application.

2.2 Thin Rectangular Section

The simplest of the thin-walled sections is the rectangle in which $b \gg t$. The section resists a St. Venant torsion T_s by developing shear stresses linearly varying in the thickness direction as shown in Fig. 2.1. The cross section is free to warp in St. Venant torsion. However, for the rectangular section and also for sections composed of thin rectangles originating from a point (such as angle, tee, cruciform and star), the contour warping will be zero. The only warping will be in the direction of the thickness called the thickness warping. For thin-walled closed sections and for open sections such as channel, zee, and I, the thickness warping is usually small and can be neglected. In these cases, the contour warping will be predominant. These ideas are further developed in the next chapter which tries to trace warping (w displacement) distributions for a variety of sections. However, it must be remembered that in St. Venant torsion considered in this chapter, the cross section will be allowed to warp freely be it the thickness warping or the contour warping.

© The Author(s), under exclusive license to Springer Nature Singapore Pte Ltd. 2022   17
K. Rajagopalan, *Torsion of Thin Walled Structures*,
https://doi.org/10.1007/978-981-16-7458-7_2

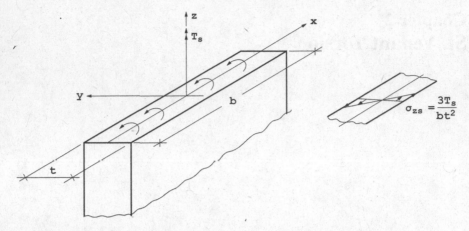

Fig. 2.1 St. Venant torsion in thin-walled rectangle

St. Venant proposed a semi-inverse method of analyzing the problem. In the semi-inverse method, part of the solution is guessed based on intuition, and the remainder of the solution is obtained so as to satisfy the differential equation of equilibrium and boundary conditions. Based on this reasoning or using Prandtl's membrane analogy, the following expression for maximum shear stress is obtained.

$$\tau_{max} = \frac{3T_s}{bt^2}$$ (2.1)

The angle of twist per unit length can be obtained as

$$\phi' = \frac{d\emptyset}{dz} = \frac{T_s}{GJ}$$ (2.2)

Here, J is the St. Venant's torsion constant given by

$$J = \frac{1}{3}bt^3$$ (2.3)

The foregoing formulas can be generalized to other thin-walled open cross sections. For curved sections such as those shown in Fig. 2.2, the above equations can be used with the understanding that b denotes the length of the curved midline of the cross section [1].

Sections such as the angle and I which are composed of several rectangles can be treated on the assumption that the total torsional moment is the sum of the torsional moments carried by each component rectangle as shown in Fig. 2.3.

In the torsional analysis, we assume that the cross section twists in its plane without any distortion in its plane, i.e., the cross section rotates as a rigid body about the center of twist. The center of twist is the point in the plane of the cross section

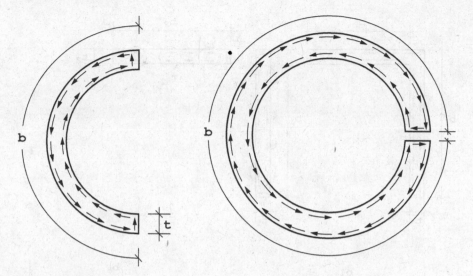

Fig. 2.2 Curved sections

which does not have any displacement due to the rotation. The whole plane of the cross section can be imagined to consist of a rigid disk rotating about an axis passing through the center of twist. The center of twist coincides with the shear center (There is, however, a distortion or deplanation of the cross section in the direction of the axis which is called warping.).

Because of the assumption that the cross section does not change its shape in its plane and rotates as a rigid-body about the center of twist, it is clear that the angle of rotation is the same for each component rectangle. Hence, we can write

$$G J_1 \phi' = T_{s1}$$

$$G J_2 \phi' = T_{s2}$$

etc., Since $T_{s1} + T_{s2} + \ldots = T_s$, we have the torsion constant of the section composed of n rectangles given by [2].

$$J = \frac{1}{3} \sum_{i=1}^{n} b_i t_i^3 \tag{2.4}$$

The shear stress in each component can be computed from Eq. (2.1) using the T_{si}, b_i and t_i of that component rectangle. It can be noted that the component having a greater thickness has the greater shearing stress.

It must be noted that the foregoing formulas apply for thin, narrow rectangles where $b \gg t$. The factor of 3 or $\frac{1}{3}$ (Eqs. 2.1 and 2.3) applies only for an infinitely long rectangle. However, in cases where b is around 10t, only small errors (some 10%

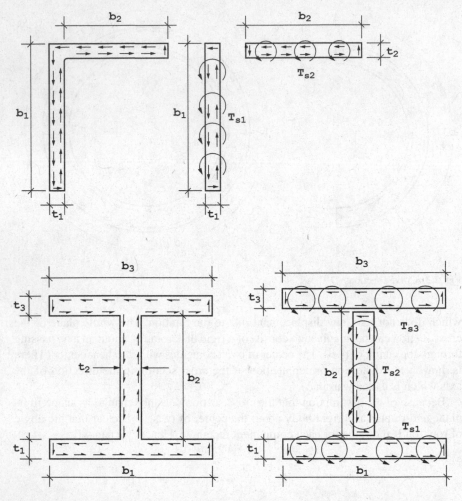

Fig. 2.3 Torsional moments as the sum of torsional moments carried by each component rectangle

for usual profiles used in engineering) are encountered in the use of the foregoing equations. The values for exact stress and rotation that can be obtained from the theory of elasticity are given in Table 2.1 for all cases of rectangles. However, a rectangle qualifies as a thin-walled structure only when $b > 10t$. We have

$$\tau_b = \text{shear stress at the midpoint of broader side}$$

$$\tau_n = \text{Shear stress at the midpoint of narrower side}$$

$$\tau_b = \frac{\alpha_1 T_s}{bt^2}$$

Table 2.1 Constants for finding exact stress and rotation

$\frac{b}{t}$	α_1	$\frac{\tau_n}{\tau_b}$	α_2
1.0	4.808	1.000	0.141
1.2	4.566	0.935	0.166
1.5	4.329	0.859	0.196
2.0	4.065	0.795	0.229
2.5	3.876	0.766	0.249
3.0	3.745	0.753	0.263
4.0	3.546	0.745	0.281
6.0	3.344	0.743	0.299
10.0	3.205	0.742	0.312
∞	3.000	0.742	0.333

$$J = \alpha_2 b t^3$$

2.3 Single Cell Thin-Walled Section

A thin-walled tube whose cross section is an arbitrarily shaped single cell has been studies by Bredt and Batho. In cellular sections, the torque is resisted by two systems of shear stresses as shown in Fig. 2.4. The first system consists of shear stresses

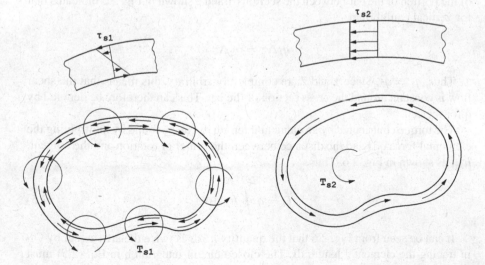

Fig. 2.4 Torsion of cellular sections

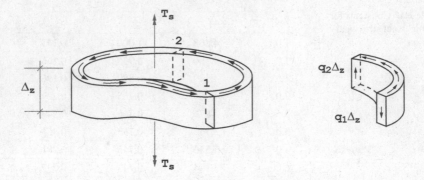

Fig. 2.5 Free body diagram of the element from the arbitrary section

linearly varying in the thickness direction as in open sections and contributes a torque T_{s1}. The second system consists of shear stresses uniformly distributed in the thickness direction constituting a shear flow in a closed loop and contributes a torque T_{s2}. The first system is comparatively insignificant compared to the second and is therefore neglected. The shear stress distribution is thus assumed to be uniformly distributed in the thickness direction with magnitude τ_{s2} although the shear stress actually varies linearly with a value of $\tau_{s2} + \tau_{s1}$ at the outer periphery and a value of $\tau_{s2} - \tau_{s1}$ at the inner periphery. The applied torque is also assumed to be entirely carried by T_{s2}.

To derive an expression for T_s, we consider a small length Δz of the thin-walled bar as shown in Fig. 2.5. If the thicknesses of the wall at two arbitrary Sects. 2.1 and 2.2 are t_1 and t_2, respectively, we can write $q_1 = \tau_1 t_1$ and $q_2 = \tau_2 t_2$. The free body of the portion of the bar between the sections 1 and 2 shown in Fig. 2.5 indicates that for vertical equilibrium

$$q_1 \Delta z = q_2 \Delta z$$

Thus, $q_1 = q_2$. Since 1 and 2 are completely arbitrary, this means that the shear flow is constant around the cross section of the bar. This can therefore be denoted by q for the cell.

The torque contributed by an elemental length ds can be found by multiplying the elemental force qds and the distance between the center of rotation and the tangent, h_{sc}, as shown in Fig. 2.6. Thus,

$$T_s = q \oint h_{sc} \mathrm{d}s \tag{2.5}$$

It can be seen from Fig. 2.6 that the quantity h_{sc}ds is twice the area swept by OP in tracing the elemental length ds. The closed circuit integration in Eq. (2.5) must therefore be equal to twice the area enclosed by the median line of the arbitrary cross section (i.e., the voided area). Hence,

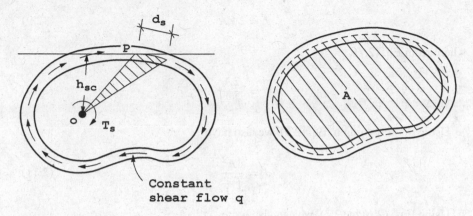

Constant
shear flow q

Fig. 2.6 Total torque

$$T_s = 2qA \tag{2.6}$$

The above formula is called the Bredt's equation.

The angle of twist can be found by a simple energy approach. Equating internal and external energies, we have

$$\frac{1}{2}T_s\phi = \oint \frac{\tau^2}{2G}(Lt)\mathrm{d}s$$

Rearranging, we get

$$\frac{\phi}{L} = \phi' = \oint \frac{(q/t)^2}{2qA}\frac{t}{G}\mathrm{d}s$$

or

$$\phi' = \frac{q}{2GA}\oint \frac{\mathrm{d}s}{t} \tag{2.7}$$

A modified shear flow ψ will be found useful. It is defined as

$$\psi = \frac{q}{G\phi'} \tag{2.8}$$

The expression for torsion constant of a single cell section is thus

$$J = \frac{T_s}{G\phi'} = \frac{2qA}{G\phi'}$$

Hence,

$$J = 2\psi A \qquad (2.9)$$

where from Eq. (2.7)

$$\psi = \frac{2A}{\oint \frac{ds}{t}} \qquad (2.10)$$

Using the foregoing in Eq. (2.9), we also have

$$J = \frac{4A^2}{\oint \frac{ds}{t}} \qquad (2.11)$$

Using Eqs. (2.6) in (2.7), we can also write

$$\phi' = \frac{T_s}{4GA^2} \oint \frac{ds}{t} \qquad (2.12)$$

The foregoing equation is called the Batho's formula.

2.4 Multicellular Section

Multicellular sections occur in ships, aircrafts, building cores and bridges. The St. Venant's torsion of multicell tubes can be analyzed by a simple extension of the one-cell analysis. A typical three-cell tube is shown in Fig. 2.7 and is used as a vehicle in deriving the general formulas for multicellular sections. The shear flow in each cell wall is constant. The constant shear flows are q_1, q_2, and q_3 for the first, second, and third cell, respectively. In the common wall, the shear flow is constant but has a value equal to the difference of the shear flows in the cells for which the wall is common. An expression for St. Venant torsional moment can be derived by taking the moment about an arbitrary point S. We have with reference to Fig. 2.8.

$$T_s = \int_B^A q_1 h_{sc} ds + \int_B^A (q_2 - q_1) h_{sc} ds + \int_D^B q_2 h_{sc} ds$$

$$+ \int_A^C q_2 h_{sc} ds + \int_C^D (q_2 - q_3) h_{sc} ds + \int_C^D q_3 h_{sc} ds$$

$$= 2q_1(A_1 + A_{21}) + 2(q_2 - q_1)A_{21} + 2q_2 A_{22}$$

$$+ 2q_2 A_{24} + 2(q_2 - q_3)A_{23} + 2q_3(A_3 + A_{23})$$

$$= 2q_1 A_1 + 2q_2 A_2 + 2q_3 A_3$$

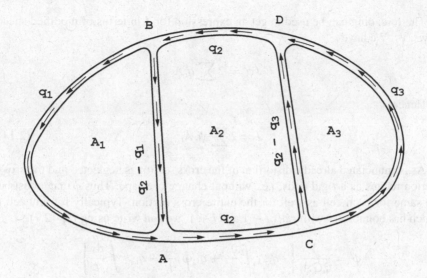

Fig. 2.7 Three-cell tube of arbitrary section

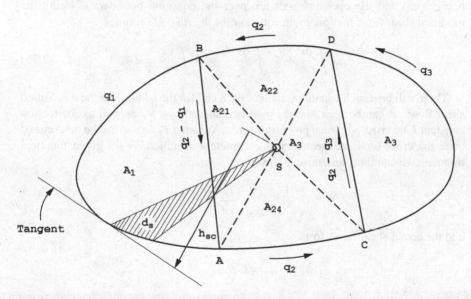

Fig. 2.8 Shear flow in three-cell tube of arbitrary section

Thus, for a multicellular section, the general formula for St. Venant torque is

$$T_s = \sum_{i=1}^{n} 2 A_i q_i \tag{2.13}$$

The foregoing can be used to get an expression for J in terms of modified shear flows, ψ_i. We have

$$GJ\phi' = 2\sum q_i A_i$$

Hence,

$$J = 2\sum \psi_i \dot{A}_i \tag{2.14}$$

As is enunciated already, distortion of the cross section is neglected and the cross section rotates as a rigid body, i.e., without change of shape. Thus ϕ or ϕ' must be the same for each cell as well for the entire cross section. Typically for any cell i which has boundaries with *cells* $i - 1$ and $i + 1$, we can write using Eq. (2.7).

$$\phi' = \frac{1}{2GA_i}\left\{-q_{i-1}\int \frac{ds}{t} + q_i \oint \frac{ds}{t} - q_{i+1}\int \frac{ds}{t}\right\}$$

where the closed integral is over the entire contour of the cell i (including the common boundaries) and the open integrals are over the common boundaries. Using the modified shear flow, the foregoing equation for the ith cell becomes

$$-\psi_{i-1}\int \frac{ds}{t} + \psi_i \oint \frac{ds}{t} - \psi_{i+1}\int \frac{ds}{t} = 2A_i \tag{2.15}$$

There will be n such equations, one for each cell for the solution of the n modified shear flows. It can be seen that the modified shear flows ψ_i as well as the torsion constant J are cross-sectional properties for St. Venant torsion and can be determined once the cross-sectional geometry is completely specified. For a given torsional moment, we can then determine ϕ' from

$$\phi' = \frac{T_s}{GJ}$$

and the actual shear flows from

$$q_i = G\phi' \psi_i$$

The foregoing formulas can be conveniently expressed in a matrix form. If we define

$$[a] = \begin{bmatrix} \oint_1 \frac{ds}{t} & -\int_{1-2} \frac{ds}{t} & & \\ -\int_{1-2} \frac{ds}{t} & \oint_2 \frac{ds}{t} & -\int_{2-3} \frac{ds}{t} & \\ & -\int_{2-3} \frac{ds}{t} & \oint_3 \frac{ds}{t} & -\int_{3-4} \frac{ds}{t} \text{ etc.,} \end{bmatrix}$$

$$\{\psi\} = \left\{ \begin{array}{c} \psi_1 \\ \psi_2 \\ \vdots \end{array} \right\} \text{ and } \{A\} = \left\{ \begin{array}{c} A_1 \\ A_2 \\ \vdots \end{array} \right\}$$

Then, we can write

$$[a]\{\psi\} = 2\{A\} \tag{2.16}$$

$$J = 2\{\psi\}^T\{A\} = \{\psi\}^T[a]\{\psi\} \tag{2.17}$$

2.5 Stress Concentration

The shear stress distribution is linear with the maximum value given by Eq. (2.1) only in the regions away from the junctions of the plate elements forming the thin-walled section. Junctions of plates produce protruding corners such as A in Fig. 2.9 (where there is 90° of material and 270° of void) and re-entrant corners such as B (where there is 270° of material and 90° of void). The shear stress distribution at junctions is nonlinear with zero at the protruding corners and maximum at the re-entrant corners.

The stress concentration $\left(\frac{\tau_c}{\tau_{max}}\right)$ will be very large when the re-entrant corners are sharp. The concentration is reduced by rounding the corners or by thickening the section. The stress concentration factors for several sectional shapes twisted under torsion are tabulated in handbooks. For angles and tees, the stress concentration factor K_S which must multiply τ_{max} to get τ_c is shown in Fig. 2.9. The factor depends on the ratio of the thickness of angle to the radius of the fillet. The stress concentration factors shown in Fig. 2.9 are obtained by Trefftz. However, more accurate stress concentration factors are obtained by Conway. The solutions of Conway are shown in Fig. 2.10 and contrasted with those of Trefftz.

The stress concentration factors at the fillet of a 90° corner in a rectangular box section under torsion are shown in Fig. 2.11 and are contrasted with the stress concentration factors for angles and tees [3].

2.6 Box with Lattice Walls

Occasionally, we may encounter closed sections with some faces constituted by a lattice framework instead of a thin plate. In these cases, the lattice face could be replaced by an equivalent thin plate. The thickness t* of the equivalent plate can be obtained by equating the strain energy in shear of the lattice face and the equivalent thin plate. Such equivalent thickness have been obtained in the literature for a few

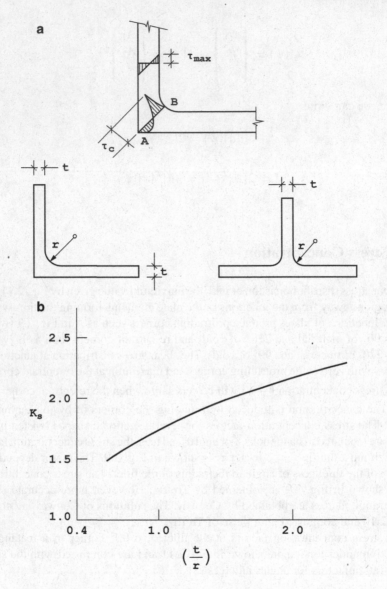

Fig. 2.9 **a** Stress concentration in tees and angles. **b** Stress concentration factor by Trefftz

frequently occurring configurations of the lattice faces. Five cases are shown in Fig. 2.12. The equivalent thicknesses are given in the following for these cases.

[a.] Here, A_1, A_2, and A_3 are the cross-sectional areas of the top chord, bottom chord, and the diagonal, respectively, and $d = \sqrt{a^2 + b^2}$

Fig. 2.10 Stress concentration factors by Conway

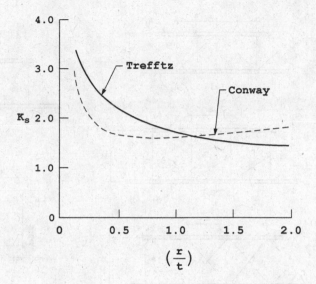

Fig. 2.11 SCF at fillet of a 90° corner in a rectangular box

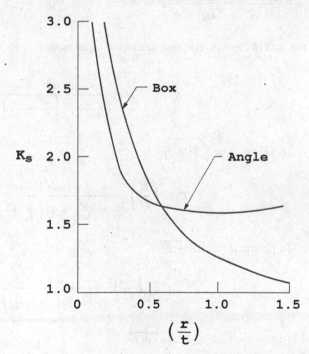

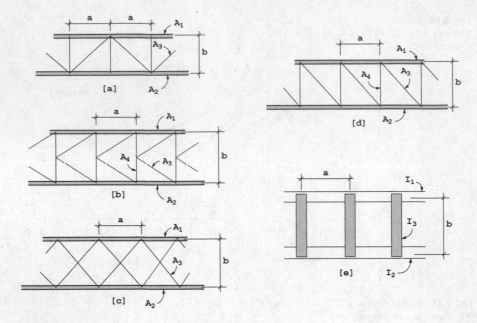

Fig. 2.12 5 Cases (**a**–**e**) of configurations of lattice faces

$$t^* = (E/G)\frac{ab}{\left(\frac{d^3}{A_3} + \frac{a^3}{3}\left[\frac{1}{A_1} + \frac{1}{A_2}\right]\right)} \tag{2.18}$$

[b.] Here, $d = \sqrt{a^2 + \frac{b^2}{4}}$

$$t^* = \left(\frac{E}{G}\right)\frac{ab}{\frac{2d^3}{A_3} + \frac{b^3}{4A_4} + \frac{a^3}{12}\left(\frac{1}{A_1} + \frac{1}{A_2}\right)} \tag{2.19}$$

[c.] Here, $d = \sqrt{a^2 + b^2}$

$$t^* = \left(\frac{E}{G}\right)\frac{ab}{\frac{d^3}{2A_3} + \frac{a^3}{12}\left(\frac{1}{A_1} + \frac{1}{A_2}\right)} \tag{2.20}$$

[d.] In this case, $d = \sqrt{a^2 + b^2}$

$$t^* = \left(\frac{E}{G}\right)\frac{ab}{\frac{d^3}{A_3} + \frac{b^3}{A_4} + \frac{a^3}{12}\left(\frac{1}{A_1} + \frac{1}{A_2}\right)} \tag{2.21}$$

[e.] Here, I_1, I_2, and I_3 are the second moment of areas of the top, bottom, and cross members

$$t^* = \left(\frac{E}{G}\right)\frac{1}{\frac{ab^2}{12I_3} + \frac{a^2b}{48}\left(\frac{1}{I_1} + \frac{1}{I_2}\right)} \qquad (2.22)$$

2.7 Composite Cross Section

In some cross sections, the constituent plates may consist of different materials. The analysis of such cross sections can be reduced to the analysis of the cross section with one reference material. Such a homogenous cross section can be obtained by considering equivalent thicknesses for the plates not made of the reference material. The multiplication of the actual thickness by the shear modular ratio, n, achieves the purpose where $n = G_{\text{plate}}/G_{\text{reference plate}}$.

As an example, the bridge cross section consisting of a concrete deck plate and a steel box shown in Fig. 2.13 is considered. The torsion constant of the section is given by Eq. (2.11). (The cantilever outstands will have negligible stiffness)

$$J = \frac{4\left[\frac{h}{2}(b + B_1)\right]^2}{\left\{\frac{2a}{t_3} + \frac{b}{t_2} + \frac{B_1}{nt_1}\right\}}$$

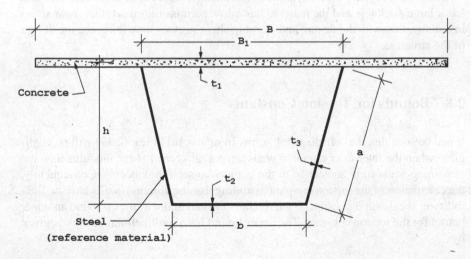

Fig. 2.13 Concrete deck plate and steel box of a bridge

$$= \frac{h^2(b + B_1)^2}{\left\{\frac{2a}{t_3} + \frac{b}{t_2} + \frac{B_1}{nt_1}\right\}} \tag{2.23}$$

In the foregoing, n is given by

$$n = \frac{G_{\text{concrete}}}{G_{\text{Steel}}}$$

A typical value of n is 0.15.
The shear flow in the box section is

$$q = \frac{T}{h(b + B_1)} \tag{2.24}$$

The shear stress in concrete deck is simply

$$\tau_c = \frac{T}{h(b + B_1)t_1} \tag{2.25}$$

The cantilever portion of the concrete deck will have a St. Venant shear stress with a maximum value of

$$\tau_{\max} = \frac{T}{J}nt_1 \tag{2.26}$$

Since the foregoing is proportional to the thickness and because the concrete slab has a large thickness and the material has a low permissible stress, this shear stress in combination with the flexure and warping shear stresses may influence the design of the structure.

2.8 Bounds for Torsion Constant

It can be seen that the 'off-diagonal' terms in matrix [a] in Eq. (2.17) will be negligible when the interior or common walls have a far greater shear modulus than the remaining walls (this is clear from the composite section concept). Consequently, the calculation of the torsion constant assuming that the common walls have no flexibility in shear will overestimate the torsion constant and will thus provide an upper bound for the torsion constant. The upper bound for a multicellular section is given by

$$J_u = 4 \sum_{i=1}^{n} \frac{A_i^2}{\sum_{j=1}^{n} a_{ij}} \tag{2.27}$$

that is, the coefficients of the matrix [a] is summed for each row and $A_i{}^2$ is divided by this sum. The sum of such calculations over all rows is found, and the upper bound is four times this value. Note that the summation of the rows in [a] is equivalent to assuming that the common walls have no flexibility in shear since the off-diagonal terms will cancel with the corresponding values included in the diagonal term.

A lower bound for the torsion constant can be computed assuming that the common walls are completely flexible in shear and so can be completely neglected. The multicellular section thus becomes only a single cell section with its wall coinciding with the outer periphery of the multicellular section. The lower bound is thus

$$J_L = \frac{4\left(\sum_{i=1}^{n} A_i\right)^2}{\sum_{i=1}^{n} \sum_{j=1}^{n} a_{ij}} \tag{2.28}$$

It is thus possible to bracket the value of the torsion constant J for any multicellular section without laborious solution of the simultaneous equations.

$$J_L < J < J_u \tag{2.29}$$

2.9 Torsion of Multicellular Section Connected by a Base Cell

Occasionally, small cells have walls in common with a larger base cell as shown in Fig. 2.14. The larger cell is denoted by zero, and the smaller cells inside the larger cells are designated by 1, 2, 3, etc. For this special arrangement, explicit expressions for St. Venant torsional properties can be derived as follows:

The equations defining the modified shear flows for this special configuration are

$$\begin{bmatrix} \oint_0 \frac{ds}{t} & -\int_{10} \frac{ds}{t} & -\int_{20} \frac{ds}{t} \\ -\int_{10} \frac{ds}{t} & \int_1 \frac{ds}{t} & 0 \\ -\int_{20} \frac{ds}{t} & 0 & \int_2 \frac{ds}{t} \end{bmatrix} \begin{Bmatrix} \psi_0 \\ \psi_1 \\ \psi_2 \end{Bmatrix} = \begin{Bmatrix} 2A_0 \\ 2A_1 \\ 2A_2 \end{Bmatrix}$$

Fig. 2.14 Multicell section connected by a base cell

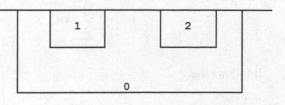

or in general for a configuration with n interior cells,

$$
\begin{bmatrix}
a_0 & -a_{10} & -a_{20} \ldots -a_{no} \\
-a_{10} & a_1 & \\
-a_{20} & & a_2 \\
\vdots & & \\
-a_{no} & & a_n
\end{bmatrix}
\begin{Bmatrix}
\psi_0 \\
\psi_1 \\
\\
\\
\psi_n
\end{Bmatrix}
=
\begin{Bmatrix}
2A_0 \\
2A_1 \\
2A_2 \\
\\
2A_n
\end{Bmatrix}
\tag{2.30}
$$

All the equations in the foregoing except the first can be written as

$$
-a_{i0}\psi_0 + a_i\psi_i = 2A_i \quad (\text{for } i = 1 \text{ to } n)
$$

Hence,

$$
\psi_i = \frac{2A_i}{a_i} + \frac{a_{i0}}{a_i}\psi_0 \tag{2.31}
$$

Thus, the modified shear flows of each interior cell is found once ψ_0 is determined. Using Eq. (2.31) in the first of Eq. (2.30), we have

$$
a_0\psi_0 - \sum_{i=1}^{n} a_{i0}\left(\frac{2A_i}{a_i} + \frac{a_{i0}}{a_i}\psi_0\right) = 2A_0
$$

which gives

$$
\psi_0\left[a_0 - \sum_{i=1}^{n} \frac{a_{i0}^2}{a_i}\right] = 2A_0 + \sum_{i=1}^{n} \frac{2A_i a_{i0}}{a_i}
$$

With the following notation

$$
\alpha = 2A_0 + \sum_{i=1}^{n} \frac{2A_i a_{i0}}{a_i}
$$

$$
\beta = a_0 - \sum_{i=1}^{n} \frac{a_{i0}^2}{a_i}
$$

$$
\gamma = \sum_{i=1}^{n} \frac{4A_i^2}{a_i} \tag{2.32}
$$

It follows that

$$
\psi_0 = \frac{\alpha}{\beta}
$$

and from Eq. (2.31)

$$\psi_i = \frac{2A_i}{a_i} + \frac{a_{i0}}{a_i}\frac{\alpha}{\beta}$$

The torsion constant can be written as

$$J = 2\psi_0 A_0 + 2\sum_{i=1}^{n} \psi_i A_i$$

$$= \left[\frac{2\alpha}{\beta}A_0 + \sum_{i=1}^{n}\frac{4A_i^2}{a_i} + \frac{\alpha}{\beta}\sum_{i=1}^{n}\frac{2A_i a_{i0}}{a_i}\right]$$

$$= \left[\frac{2\alpha}{\beta}A_0 + \gamma + \frac{\alpha}{\beta}(\alpha - 2A_0)\right]$$

Hence,

$$J = \gamma + \frac{\alpha^2}{\beta} \tag{2.33}$$

The actual shear flows can be determined as

$$q_0 = \psi_0 G\phi' = \frac{T}{J}\psi_0 = \frac{T\beta}{\alpha^2 + \gamma\beta}\frac{\alpha}{\beta}$$

Hence,

$$q_0 = \frac{T\alpha}{\alpha^2 + \gamma\beta} \tag{2.34}$$

Also

$$q_i = \frac{T}{J}\psi_i$$

and

$$q_{i0} = q_0 - q_i$$

This completes the solution of the multicellular section. For this special multicell case, the torsional properties are explicitly represented by Eqs. (2.33) and (2.34).

Fig. 2.15 Review problem 1

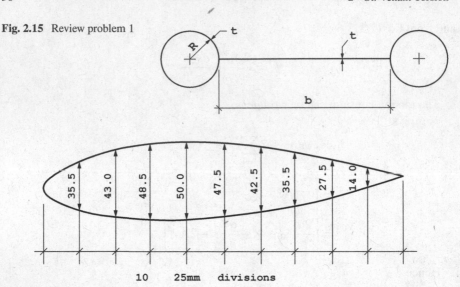

Fig. 2.16 Review problem 2

2.10 Review Problems

1. What must be the value of R in Fig. 2.15 if the torsional stiffness of the middle narrow rectangular section is to be doubled by adding the two closed thin-walled circular tubes?
2. A propeller blade has the cross section shown in Fig. 2.16. Calculate the maximum shear stress for a given twisting moment T.
3. Is there any error in the calculated shear flows shown in Fig. 2.17?
4. A torque of 2 kNm is applied to a three-cell tube shown in Fig. 2.18. If the fillet radii at all re-entrant corners is 0.8 mm, find the maximum shear stress.
5. Compute the torsion constant of the section shown in Fig. 2.19. Also show that the value can be bracketed between two bounds.
6. Determine the torsion constant of the multicellular section shown in Fig. 2.20.

2.11 Answers to Review Problems

1. We have

$$\frac{1}{3}bt^3 + 4\pi R^3 t = 2\left[\frac{1}{3}bt^3\right]$$

Hence,

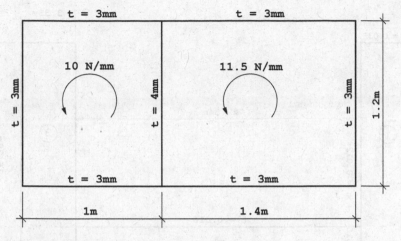

Fig. 2.17 Review problem 3

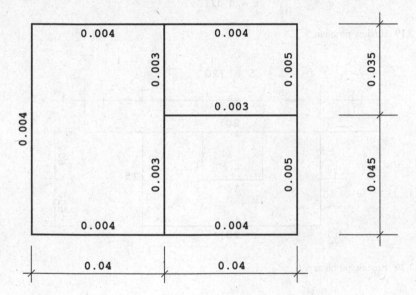

Fig. 2.18 Review problem 4

$$R = \left(\frac{bt^2}{12\pi}\right)^{\frac{1}{3}}$$

2. The thin rectangular section formula for torsion constant can be applied to compute J. Hence,

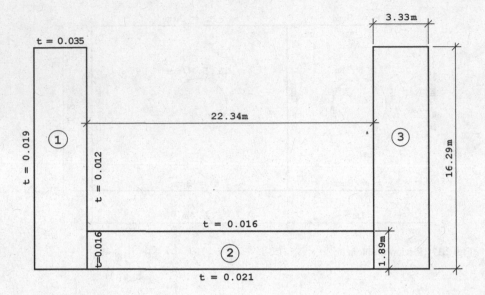

Fig. 2.19 Review problem 5

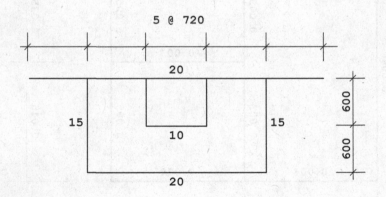

Fig. 2.20 Review problem 6

$$J = \frac{1}{3} \int [dx] t^3$$

The integral can be evaluation by Simpson's numerical integration. We have

Odd ordinates	Cube	Even ordinates	Cube
43.0	79,507	35.5	44,738.875
50.0	125,000	48.5	114,084.120
42.5	76,765.625	47.5	107,171.870

(continued)

(continued)

Odd ordinates	Cube	Even ordinates	Cube
27.5	20,796.875	35.5	44,738.875
		14.0	2744.000
Sum	**302,069.500**	**Sum**	**313,477.75**

Hence,

$$J = \frac{1}{3}\left[\frac{25}{3}\{0 + 0 + 2[302069.5] + 4[313477.75]\}\right]$$

$$= 5,161,250 \, \text{mm}^4$$

The maximum shear stress is

$$\tau_{max} = \frac{T}{5161250}(50) = 9.6876 \times 10^{-6}T$$

3.   The solution for the modified shear flows can be written as

$$-\left(\frac{1200}{4}\right)\psi_2 + \left[\frac{1000}{3} + \frac{1000}{3} + \frac{1200}{3} + \frac{1200}{4}\right]\psi_1 = 2(1000)(1200)$$

$$\left[\frac{1400}{3} + \frac{1400}{3} + \frac{1200}{3} + \frac{1200}{4}\right]\psi_2 - \frac{1200}{4}\psi_1 = 2(1400)(1200)$$

Solving

$$\psi_1 = 3745.2548 \text{ and } \psi_2 = 2732.1162$$

The torsion constant is

$$J = 2\psi_1 A_1 + 2\psi_2 A_2$$

$$= 2(3745.2548)(12 \times 10^5) + 2(2732.1162)(16.8 \times 10^5)$$

$$= 1.81685 \times 10^{10} \, \text{mm}^4$$

The shear flows are

$$q_1 = G\phi'\psi_1 = \frac{T}{J}\psi_1 = 2061.4001 \times 10^{-10}T$$

$$q_2 = 1503.7654 \times 10^{-10}T$$

Thus,

$$\frac{q_1}{q_2} = 1.3708$$

But the calculated shear flows are in the ratio

$$\frac{10}{11.5} = 0.8696$$

Hence, there is an error. If q_1 is correct, q_2 must be 7.3 N/mm. These correspond to a torque

$T = 0.0485$ MNm

4.   The equations for the modified shear flows are

$$\left[2\frac{0.04}{0.004} + \frac{0.08}{0.004} + \frac{0.08}{0.003}\right]\psi_1 - \frac{0.035}{0.003}\psi_2 - \frac{0.045}{0.003}\psi_3 = 2[0.0032]$$

$$-\frac{0.035}{0.003}\psi_1 + \left[\frac{0.04}{0.003} + \frac{0.04}{0.004} + \frac{0.035}{0.005} + \frac{0.035}{0.003}+\right]\psi_2 - \frac{0.04}{0.003}\psi_3 = 2[0.0014]$$

$$-\frac{0.045}{0.003}\psi_1 - \frac{0.04}{0.003}\psi_2 + \left[\frac{0.04}{0.004} + \frac{0.04}{0.003} + \frac{0.045}{0.005} + \frac{0.045}{0.003}\right]\psi_3 = 2[0.0018]$$

Solving we get

$$\psi_1 = 1.6499479 \times 10^{-4}$$

$$\psi_2 = 1.6829209 \times 10^{-4}$$

$$\psi_3 = 1.7574966 \times 10^{-4}$$

Also

$$J = 2[10^{-4}]\{(1.6499479)(0.0032) + (1.6829209)(0.0014)$$
$$+(1.7574966)(0.0018)\} = 0.0215988 \times 10^{-4} \text{ m}^4$$

The shear flows are

$$q_1 = \frac{T}{J}\psi_1 = \frac{2(10^{-3})(1.6499479 \times 10^{-4})}{0.0215988 \times 10^{-4}}$$
$$= 0.1527814 \text{ MN/m}$$

$$q_2 = 0.1558346 \text{ MN/m}$$

$$q_3 = 0.1627402 \text{ MN/m}$$

These shear flows are nearly the same. Hence, the internal plates do not carry much shear stress. The maximum shear stress must therefore occur at one of the outer walls with thickness of 4 mm. Hence, this must occur in the outer wall of cell 3 where the shear flow is maximum. Thus, the maximum nominal shear stress is

$$\tau_{max} = \frac{0.1627402}{0.004} = 40.685052\,\text{MPa}$$

The stress concentration factor in torsion at the re-entrant corner of a box section with $(r/t) = (0.8/4) = 0.2$ can be obtained from Fig. 2.11 as $K_s = 2.82$. Hence, the absolute maximum shear stress

$$\tau_C = 2.82(40.685052) = 115\,\text{MPa}$$

5. The equations for the modified shear flows are

$$\left[\frac{3.33}{0.035} + \frac{3.33}{0.021} + \frac{16.29}{0.019} + \frac{14.4}{0.012} + \frac{1.89}{0.016}\right]\psi_1 - \frac{1.89}{0.016}\psi_2$$
$$= 2(16.29)(3.33)$$

$$-\frac{1.89}{0.016}\psi_1 + \left[\frac{22.34}{0.016} + \frac{22.34}{0.021} + 2\frac{1.89}{0.016}\right]\psi_2 - \frac{1.89}{0.016}\psi_3$$
$$= 2(22.34)(1.89)$$

$$-\frac{1.89}{0.016}\psi_2 + \left[\frac{3.33}{0.035} + \frac{3.33}{0.021} + \frac{16.29}{0.019} + \frac{14.4}{0.012} + \frac{1.89}{0.016}\right]\psi_3$$
$$= 2(16.29)(3.33)$$

Solving we get

$$\psi_1 = 0.0464$$

$$\psi_2 = 0.0354$$

$$\psi_3 = 0.0464$$

The torsion constant is

$$J = 2\sum \psi_i A_i$$
$$= 2(2)(0.0464)(16.29)(3.33) + 2(0.0354)(22.34)(1.89)$$
$$= 13.057362\,\text{m}^4$$

Using Eq. (2.28), we have

$$J_\text{L} = \frac{4(150.714)^2}{7082.2249} = 12.8291 \, \text{m}^4$$

Also using Eq. (2.27), we get

$$J_\text{u} = 4\left[\frac{(54.2457)^2}{2311.0827}(2) + \frac{(42.2226)^2}{2460.0595}\right] = 13.0847$$

Thus,

$$12.8291 < 13.0574 < 13.0847$$

The torsion constant is closer to the upper bound because the common walls being short are comparatively rigid.

6. For this problem, equations developed in Sect. 2.9 can be used. We have

$$a_0 = \frac{1200}{15} + \frac{1200}{15} + \frac{2160}{20} + \frac{2160}{20} = 376$$

$$a_{10} = \frac{720}{20} = 36$$

$$a_1 = \frac{720}{10} + \frac{720}{20} + \frac{600}{10} + \frac{600}{10} = 228$$

$$A_0 = (2160)(1200) = 2592000$$

$$A_1 = (720)(600) = 432000$$

$$\alpha = 2(2592000) + \frac{2(432000)(36)}{228} = 5320421$$

$$\beta = 376 - \frac{(36)^2}{228} = 370.31579$$

$$\gamma = \frac{4(432000)^2}{228} = 3.2741053 \times 10^9$$

Thus,

$$J = 0.003(10^{12}) + \frac{(5.320421)^2(10^{12})}{370.31579} = 0.08 \times 10^{12} \, \text{mm}^4$$

References

1. Oden, J.T., Ripperger, E.A.: Mechanics of Elastic Structures. McGraw Hill, New York (1981)
2. Kollbrunner, C.F., Basler, K.: Torsion in Structures. Springer Verlag, Berlin (1969)
3. Rajagopalan, K.: Comment on "A simple formula for the maximum stress in a twisted angle or channel." Int. J. Mech. Sci. **15**, 775–776 (1973)

Chapter 3
Warping Properties of Thin-Walled Sections

3.1 Introduction

Thin-walled sections may be subjected to mixed torsion or combined bending and torsion. The analysis of these problems becomes easy when a series of warping properties based on the geometry of the thin-walled section are defined. The evaluation of these properties is rendered easier by making use of sectorial coordinates. In this chapter, several sectional properties are defined and evaluated. The physical meaning of some of them is explained briefly but their use would be further illustrated in the succeeding chapters.

3.2 Sectorial Area

The warping displacement w of the contour (middle thickness line) of the thin-walled cross section is proportional to the rate of twist in St. Venant torsion. Often this fact is continued even in mixed torsion situations. Vlasov made the following assumption [1]:

$$w(s, z) = -\phi'(z)\omega(s) \tag{3.1}$$

In the foregoing, w is the out-of-plane warping displacement of the cross section at point s in the contour and $\omega(s)$ is the 'unit warping.' It is easily seen that when the rate of twist is $\phi' = -1$, the warping displacement is $\omega(s)$. Hence, $\omega(s)$ is called the unit warping. For any other value of rate of twist, $\omega(s)$ represents the warping displacement to some scale. The unit warping can be seen to have the dimension m^2 from Eq. (3.1). Having established that the warping function represents an area, it will be shown how this area is defined and evaluated.

The definition of sectorial area is shown in Fig. 3.1. At a typical point i, the radius vector pi sweeps the shaded area over a distance ds along the contour. This area is

© The Author(s), under exclusive license to Springer Nature Singapore Pte Ltd. 2022
K. Rajagopalan, *Torsion of Thin Walled Structures*,
https://doi.org/10.1007/978-981-16-7458-7_3

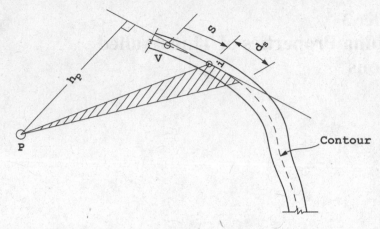

Fig. 3.1 Sectorial area

equal to $\frac{1}{2}h_p\mathrm{d}s$, where h_p is the tangential distance from the point P at the point i. The increment of unit warping at i is defined as

$\mathrm{d}\omega(s) = h_p\mathrm{d}s$ = twice the sectorial area swept over $\mathrm{d}s$

The area will be positive if it is swept in the positive direction of s. In the foregoing P is called the pole about which the sectorial area is traced. The unit warping at the point i depends on the pole and the initial point V on the contour where the warping is zero. The point V is called the sectorial origin. Thus,

$$\omega(s) = \int_0^s h_p\mathrm{d}s \qquad (3.2)$$

which will of course be equal to twice the area swept from V to i. It can be seen that if the integration proceeds around a closed section $\omega = 2A$ where A is the (*voided*) area enclosed by the medial line. This is true whatever may be the locations of the pole and the sectorial origin.

3.3 Integrals Involving Sectorial Area

Integrals involving sectorial area are needed in the definition of warping properties. Some of these are [2]

$$\text{warping static moment} \quad \int \omega \mathrm{d}A_s = \int \omega t \mathrm{d}s$$

$$\text{Warping linear moments} \quad \int \omega x \mathrm{d}A_s \text{ and } \int \omega y \mathrm{d}A_s$$

$$\text{Warping moment of inertia} \quad \int \omega^2 dA_s$$

In the forgoing dA_s denotes the incremental material area in the cross section. The symbol A_s is used to denote the material area since the symbol A has been used to denote the voided area enclosed by the medial line of a cell.

In the evaluation of the integrals, a method attributed to Vereshchagin is useful. This method gives a formula for the integral evaluation where the integrand involves the product of two functions of s, one nonlinear and the other linear. As shown in Fig. 3.2, we have

$$\int_0^L f_1(s)f_2(s)ds = \int_0^L f_1(s)[a+bs]ds$$

$$= a\int_0^L f_1(s)ds + b\int_0^L f_1(s)sds$$

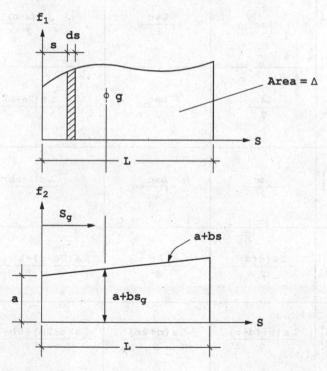

Fig. 3.2 Evaluation of Integral [Vereshchagin]

The second integral on the right-hand side locates the centroidal distance of the area under the f_1 curve. Thus,

$$\int_0^L f_1 f_2 ds = a\Delta + b\Delta s_g = \Delta(a + bs_g)$$

$$= \text{(area under } f_1)\text{(value of } f_2 \text{ at the centroidal distance of area under } f_1)$$

This is Vereshchagin's theorem. For a few standard shapes of f_1 and f_2 curves, the value of the product integrals is given in Fig. 3.3. They may be verified using Vereshchagin's principle.

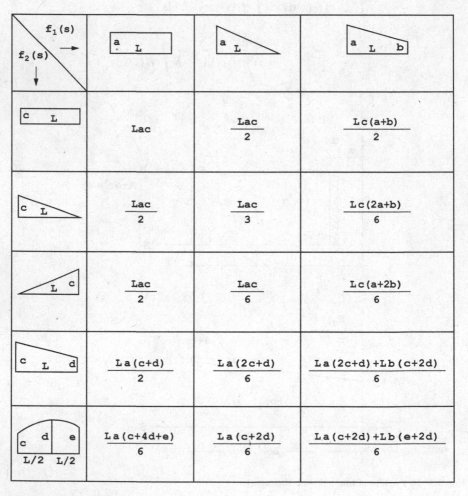

Fig. 3.3 Product Integral

3.4 Principal Sectorial Area

When ω is constructed with the pole located at the shear center, then the warping linear moments vanish. This is true whatever may be the location of the initial point V. If in addition to the vanishing of warping linear moments (*due to the pole taken at the shear centre*), the warping static moment also vanishes, the sectorial area will be called the principal sectorial area. The unit warping associated with the principal sectorial area is called the normalized unit warping. The normalized unit warping is obtained by a suitable choice of the initial point. It indicates a balance of warping displacements, i.e., positive and negative (into the plane and out of the plane) so that the average warping is zero. It has been pointed out in the first chapter that for the decoupling of the problems of bending and torsion, principal axes and principal sectorial area must be used. This idea will be further developed in detail in the next chapter.

To evaluate the principal sectorial area, an arbitrary initial point may be selected and ω is evaluated with the pole located at the shear center. A constant value of ω_0 is now subtracted from ω, so that $\int \omega_n dA_s = 0$. Thus, the normalized unit warping is given by

$$\omega_n = \omega - \omega_0$$

where to satisfy the condition $\int \omega_n dA_s = 0$, we must have

$$\int \omega dA_s - \int \omega_0 dA_s = 0$$

or

$$\omega_0 = \frac{\int \omega dA_s}{A_s} \qquad (3.3)$$

where ω is the warping with an arbitrary initial point and the pole at the shear center and A_s is the area of cross section of the thin-walled section. The normalized unit warping will still have the shear center as its pole (as subtraction of a constant value does not affect the location of the pole). The location of the initial point is, however, automatically adjusted in this process.

3.5 Transformations of Unit Warping

Often it may be required to find the unit warping with respect to a new pole M given the unit warping with respect to an existing pole L, the initial point being to same in both the cases. To express the transformation, the following general formula for incremental unit warping $d\omega_L$ will be useful. Referring to Fig. 3.4, we have

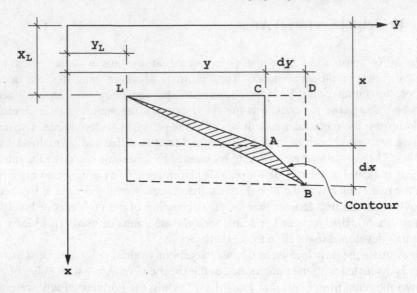

Fig. 3.4   Transformation of warping

$$d\omega_L = 2x\,\text{areaLAB}$$
$$= [\text{areaLDB} - \text{areaLCA} - \text{areaABCD}]$$
$$= 2\left\{\frac{1}{2}(y - y_L + dy)(x - x_L + dx)\right.$$
$$\left. - \frac{1}{2}(x - x_L)(y - y_L) - \frac{dy}{2}[x - x_L + dx + x - x_L]\right\}$$

Neglecting product of infinitesimal quantities, we have

$$d\omega_L = (y - y_L)(x - x_L) + (y - y_L)dx + (x - x_L)dy$$
$$- (x - x_L)(y - y_L) - (x - x_L)dy - (x - x_L)dy$$
$$= (y - y_L)dx - (x - x_L)dy \tag{3.4}$$

The foregoing formula is known as the Liebnitz formula. It can be used to transform the unit warping from one pole to the other, the initial point remaining the same through the transformation. We have, referring to Fig. 3.5,

$$d\omega_L = (y - y_L)dx - (x - x_L)dy$$

$$d\omega_M = (y - y_M)dx - (x - x_M)dy$$

Hence,

Fig. 3.5 Transformation of
unit warping

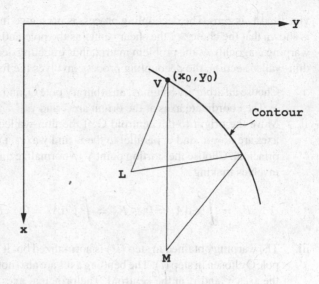

$$d[\omega_L - \omega_M] = (y_M - y_L)dx - (x_M - x_L)dy$$

Integrating we find

$$\omega_L = \omega_M + (y_M - y_L)x - (x_M - x_L)y + \Delta_\omega$$

where Δ_ω is an integration constant which can be evaluated by equating the values
of ω_L and ω_M at the starting point

$$\omega_M + (y_M - y_L)x_0 - (x_M - x_L)y_0 + \Delta_\omega = \omega_M$$

Hence,

$$\Delta_\omega = (x_M - x_L)y_0 - (y_M - y_L)x_0$$

Hence,

$$\omega_L = \omega_M + (y_M - y_L)(x - x_0) - (x_M - x_L)(y - y_0) \qquad (3.5)$$

3.6   Decoupling Equations

It has been said that for the decoupling of bending and torsion problems, the pole
is taken at the shear center. This means that $\int \omega x dA_s$ and $\int \omega y dA_s$ must vanish. In
addition, the initial point must also be chosen so that the unit warping is normalized,

viz. $\int \omega dA_s$ is zero. The decoupling process is presented in the next chapter where it is shown that the choice of the shear center as the pole and the normalization of unit warping diagonalizes the problem matrix, thus effecting decoupling. For an arbitrary thin-walled section, the decoupling process involves the following steps.

i.   Choose an arbitrary origin A, an arbitrary pole Q, and an arbitrary starting point V. The coordinate axes at the origin are x and y.

ii.   Move the origin to the centroid C of the thin-walled section. The coordinate axes are now x' and y' parallel to the x- and y-axes, i.e., they are not still made principal. Choose the starting point V to normalize unit warping. This step thus involves making

$$F_{x'} = \int x' dA_s = 0; \quad F_{y'} = \int y' dA_s = 0; \quad F_{\omega'} = \int \omega' dA_s = 0$$

iii. The warping obtained in step (ii) is normalized but it still refers to the arbitrary pole Q chosen in step (i). The bending axes are also not principal. Hence, incline the axes x' and y' at the centroid. The principal axes are denoted by X and Y. The angle of inclination can be chosen to make the product of inertia vanish, i.e.,

$$F_{XY} = \int XY dA_s = 0$$

The pole is finally moved to the shear center S, whose location can be determined my making

$$F_{X\omega_n} = \int X\omega_n dA_s = 0$$

$$F_{Y\omega_n} = \int Y\omega_n dA_s = 0$$

The normalized unit warping with respect to the shear center can be found using transformation equations.

Although the foregoing procedure is useful for arbitrary thin-walled sections, it is not necessary to follow this procedure for the decoupling process in cases of simple thin-walled sections with an axis of symmetry. Normally, the shear center of these sections is available located by some other process such as the equilibrium method described in the appendix. Also, the starting point of zero warping can be chosen as the point of intersection of the axis of symmetry with the section contour. The normalized unit warping can therefore be written down directly using the shear center as the pole and this point as the sectorial origin. However, for complex thin-walled sections, the foregoing procedure has to be followed and the necessary equations encountered in the various steps are presented in the following.

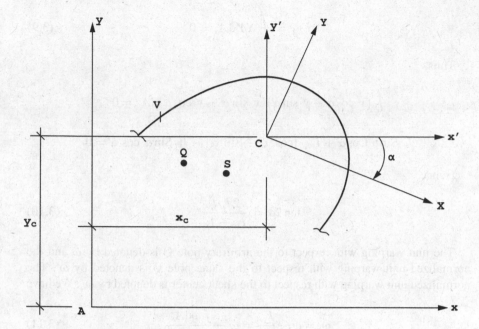

Fig. 3.6 Normalized Warping sectorial coordinates

In Fig. 3.6, the x- and y-axes are located at any arbitrary origin A and Q and V are arbitrary pole and sectorial origin, respectively. The location of the centroid C can be found as

$$x_c = \frac{F_x}{A_s} = \frac{\int x \, dA_s}{A_s}$$

$$y_c = \frac{F_y}{A_s} = \frac{\int y \, dA_s}{A_s} \tag{3.7}$$

The axes at the centroid parallel to the x- and y-axes are called x' and y'-axes. These axes can be rotated through an angle α as shown in Fig. 3.6 to become the principal bending axes. These axes are shown as X and Y. The z-axis is normal to the plane of the section and so is not transformed in any of these operations. We can write

$$X = x' \cos\alpha - y' \sin\alpha$$

$$Y = x' \sin\alpha + y' \cos\alpha \tag{3.8}$$

The principal axis inclination is found from the condition that the product of inertia vanishes.

$$F_{XY} = \int XY dA_s = 0 \tag{3.9}$$

Thus,

$$\int (x' \cos\alpha - y' \sin\alpha)(x' \sin\alpha + y' \cos\alpha) dA_s = 0$$

$$I_{y'} \sin\alpha \cos\alpha + I_{x'y'} (\cos^2\alpha - \sin^2\alpha) - I_{x'} \sin\alpha \cos\alpha = 0$$

Hence,

$$\tan 2\alpha = \frac{2I_{x'y'}}{I_{x'} - I_{y'}} \tag{3.10}$$

The unit warping with respect to the arbitrary pole Q is denoted by ω and the normalized unit warping with respect to the same pole Q is denoted by ω'. The normalized unit warping with respect to the shear center is denoted by ω_n. We have

$$\omega' = \omega - \frac{F_\omega}{A_s} = \omega - \frac{\int \omega dA_s}{A_s} \tag{3.11}$$

Hence, using the transformation Eq. (3.5), we have

$$\omega_n = \omega' + \left(y'_Q - y'_S\right)\left(x' - x'_0\right) - \left(x'_Q - x'_S\right)\left(y' - y'_0\right)$$

where $\left(x'_0, y'_0\right)$ is the initial point. However, we must have

$$\int y' dA_s = \int x' dA_s = \int \omega' dA_s = 0 \tag{3.12}$$

The shear center must be located on the conditions that $\int \omega_n X dA_s = \int \omega_n Y dA_s = 0$. Thus,

$$\int \omega_n X dA_s = \int \left[\omega' + \left(y'_Q - y'_S\right)\left(x' - x'_0\right) - \left(x'_Q - x'_S\right)\left(y' - y'_0\right)\right]$$

$$[(x' \cos\alpha - y' \sin\alpha)] dA_s = 0$$

$$F_{\omega'x'} \cos\alpha - F_{\omega'y'} \sin\alpha + \left(y'_Q - y'_S\right)I_{y'} \cos\alpha - \left(y'_Q - y'_S\right)I_{x'y'} \sin\alpha$$
$$- \left(x'_Q - x'_S\right)I_{x'y'} \cos\alpha + \left(x'_Q - x'_S\right)I_{x'} \sin\alpha = 0 \tag{3.13}$$

Also,

$$\int \omega_n Y dA_s = \int \left[\omega' + ((y'_Q - y'_S)(x' - x'_0) - (x'_Q - x'_S)(y' - y'_0)) \right]$$
$$\left[(x' \sin\alpha + y' \cos\alpha) \right] dA_s = 0$$
$$F_{\omega'x'} \sin\alpha + F_{\omega'y'} \cos\alpha + (y'_Q - y'_S) I_{y'} \sin\alpha + (y'_Q - y'_S) I_{x'y'} \cos\alpha$$
$$- (x'_Q - x'_S) I_{x'y'} \sin\alpha - (x'_Q - x'_S) I_{x'} \cos\alpha = 0 \qquad (3.14)$$

We simplify Eqs. (3.13) and (3.14) by multiplying them by $\cos\alpha$ and $\sin\alpha$ and again by $-\sin\alpha$ and $\cos\alpha$, respectively, and add the resulting equations in each case. We get

$$F_{\omega'x'} + \left(y'_Q - y'_S\right) I_{y'} - \left(x'_Q - x'_S\right) I_{x'y'} = 0$$

$$F_{\omega'y'} + \left(y'_Q - y'_S\right) I_{x'y'} - \left(x'_Q - x'_S\right) I_{x'} = 0$$

The coordinates of the shear center can be solved from the foregoing equations

$$x'_S = x'_Q + \frac{F_{\omega'x'} I_{x'y'} - F_{\omega'y'} I_{y'}}{I_{x'} I_{y'} - I_{x'y'}^2} \qquad (3.15)$$

$$y'_S = y'_Q + \frac{F_{\omega'x'} I_{x'} - F_{\omega'y'} I_{x'y'}}{I_{x'} I_{y'} - I_{x'y'}^2} \qquad (3.16)$$

It must be noted that the initial point $\left(x'_0, y'_0\right)$ does not have an effect on the foregoing integrals involving ω_n since the origin of the coordinates is at the centroid of the thin-walled section. It will have no effect even on $\int \omega_n dA_s$ which can be made equal to zero by adjusting the initial point. Thus the normalized unit warping can be simply written as

$$\omega_n = \omega' + \left(y'_Q - y'_S\right) x' - \left(x'_Q - x'_S\right) y' \qquad (3.17)$$

The foregoing clearly shows that $\int \omega_n dA_s$ is zero because of Eq. (3.12). It may be noted that the initial point is automatically adjusted in this process.

The principal moments of inertias of the thin-walled cross section are of interest. We have

$$I_X = \int Y^2 dA_s$$

$$= \int \left(x' \sin\alpha + y' \cos\alpha\right)^2 dA_s$$

$$= I_{y'} \sin^2\alpha + I_{x'y'} \sin 2\alpha + I_{x'} \cos^2\alpha$$

$$= \frac{I_{x'} + I_{y'}}{2} \pm \frac{1}{2} \sqrt{\left(I_{x'} - I_{y'}\right)^2 + 4 I_{x'y'}^2} \qquad (3.18)$$

Similarly we have

$$I_Y = \frac{I_{x'} + I_{y'}}{2} \mp \frac{1}{2}\sqrt{\left(I_{x'} - I_{y'}\right)^2 + 4I_{x'y'}^2} \tag{3.19}$$

In the foregoing formulae, the $+$ sign is used for the major principal axis and the—sign for the minor principal axis.

The warping moment of inertia is singularly important in mixed torsion calculations

$$I_\Omega = \int \omega_n^2 dA_s$$

$$= \int \left[\omega' + (y_Q' - y_S')x' - (x_Q' - x_S')y'\right]^2 dA_s$$

$$= \int \omega'^2 dA_s + (y_Q' - y_S')^2 \int x'^2 dA_s + (x_Q' - x_S')^2 \int y'^2 dA_s$$

$$+ 2(y_Q' - y_S') \int \omega'x' dA_s - 2(x_Q' - x_S')(y_Q' - y_S') \int x'y' dA_s$$

$$- 2(x_Q' - x_S') \int \omega'y' dA_s$$

The foregoing relation can be simplified by using the following already known facts.

$$\int \omega_n X dA_s = \int \omega_n Y dA_s = 0$$

Thus,

$$\int \left[\omega' + \left(y_Q' - y_S'\right)x' - \left(x_Q' - x_S'\right)y'\right][x'\cos\alpha - y'\sin\alpha] dA_s = 0$$

$$\int \left[\omega' + \left(y_Q' - y_S'\right)x' - \left(x_Q' - x_S'\right)y'\right][x'\sin\alpha + y'\cos\alpha] dA_s = 0$$

Multiplying the first by $\sin\alpha$ and the second by $\cos\alpha$ and subtracting we have

$$\int \left[\omega'y' + \left(y_Q' - y_S'\right)x'y' - \left(x_Q' - x_S'\right)y'^2\right] dA_s = 0$$

Multiplying the first by $\cos\alpha$ and the second by $\sin\alpha$ and adding we have

$$\int \left[\omega'x' + \left(y_Q' - y_S'\right)x'^2 - \left(x_Q' - x_S'\right)x'y'\right] dA_s = 0$$

Now the expression for warping moment of inertia can be arranged as

$$I_\Omega = \int \omega'^2 dA_s + (y'_Q - y'_S) \int \omega' x' dA_s - (x'_Q - x'_S) \int \omega' y' dA_s$$

$$+ (y'_Q - y'_S) \int [\omega' x' + (y'_Q - y'_S)x'^2 - (x'_Q - x'_S)x'y'] dA_s$$

$$- (x'_Q - x'_S) \int [\omega' y' + (y'_Q - y'_S)x'y' - (x'_Q - x'_S)y'^2] dA_s$$

The last two integral are zero as derived in the foregoing.
Thus,

$$I_\Omega = I_{\omega'} + \left[\frac{F_{\omega'x'}[F_{\omega'y'}I_{x'y'} - F_{\omega'x'}I_{x'}] - F_{\omega'y'}[F_{\omega'y'}I_{y'} - F_{\omega'x'}I_{x'y'}]}{I_{x'}I_{y'} - I_{x'y'}^2} \right] \tag{3.20}$$

3.7 Contour Warping of Multicell Sections

Equation (3.2) for sectorial area applies only to open sections. The warping of closed sections is less compared to the warping of open sections. This is because of the constant shear flow ψ_i (*evaluated in the previous chapter*) opposing the tendency of the section to warp. The unit warping of a typical cell in a multicellular section is thus given by

$$\omega(s) = \int \left[h_{sc} - \frac{\psi_i}{t} \right] ds \tag{3.21}$$

where h_{sc} is the tangential distance of element ds from the shear center. Except for this new formula for the computation of unit warping, all the formulae derived in the preceding sections apply for multicellular sections. It must be noted that the integrals for unit warping must use the fact that the unit warping must be the same (*unique*) at a node where a number of plates are joined together. It is also clear that Eq. (3.21) applies for open sections also for which $\psi_i = 0$.

3.8 Thickness Warping

The warping in a thin-walled section is mainly the warping of the contour, i.e., the middle surface line of the plates constituting the thin-walled section. The wall also warps relative to the contour in the thickness direction, and hence, the total warping must be given by the contour warping and the thickness warping. The thickness warping is normally negligible although in some cases it may be dominant. The foregoing treatment shows that contour warping depends on the tangential distance

h_{sc} and hence for open sections consisting of plates which originate from a point (which is the shear center), the contour warping is zero. Thus, for angles, tees, etc., the contour warping is zero and the warping consists only of thickness warping. The thickness warping at a point distance n from middle surface is given by

$$\omega_t(n, s) = -nq(s) \tag{3.22}$$

where q is the normal distance from the shear center, i.e., the distance of the normal at the section s from the shear center. At the middle surface ($n = 0$), the thickness warping is zero. The thickness warping becomes important in open sections such as the angles and tees as it is the only warping present in these sections, the contour warping being zero.

3.9 Sectorial Shear Function

It has been described earlier that when warping displacements are restrained, a warping torsional moment develops which gives rise to the warping shear stress distribution. The sectorial shear function S_Ω describes the warping shear stress distribution as shown in Eq. (1.2) which, however, applies for open sections only. Here

$$S_\Omega = \int_0^s \omega_n \mathrm{d}A_s \tag{3.23}$$

where the integration must be started at a free edge.

In closed sections, imaginary cuts are made and the integration for each cell started at these cuts. The dislocations in the cells must be corrected by introducing a constant shear flow q_w one for each cell. Thus for each cell

$$\oint (\tau_{wO} + \tau_W)\mathrm{d}s \doteq 0$$

or

$$\oint \left(\tau_{wO} + \frac{q_w}{t}\right)\mathrm{d}s = 0 \tag{3.24}$$

The total shear flow due to warping is thus [3]

$$q = q_W - \frac{T_\Omega S_{\Omega 0}}{I_\Omega} \tag{3.25}$$

where $S_{\Omega 0}$ is the sectorial shear function in the section with imaginary cuts (the ω_n used in the calculation S is the principal sectorial area found for the uncut section with respect to the shear center S of the uncut section. The fact that the shear center of the cut section must be different need not be taken notice of since the cuts are imaginary and compatibility will be restored by q_W). The compatibility condition Eq. (3.24) gives

$$[a]\{q_W\} = \{Q_{W0}\} \tag{3.26}$$

where $[a]$ is the same matrix as given in Eq. (2.16) and

$$\{q_W\} = \begin{Bmatrix} q_{W1} \\ q_{W2} \\ . \\ . \end{Bmatrix} \text{ and } \{Q_{W0}\} = \frac{T_\Omega}{I_\Omega} \begin{Bmatrix} \oint_1 \frac{S_{\Omega 0} ds}{t} \\ \oint_2 \frac{S_{\Omega 0} ds}{t} \\ . \\ . \end{Bmatrix}$$

The total shear stress including the St. Venant's torsional shear stress can also be written using a single sectorial shear function S_ω

$$\tau = \frac{T_\Omega}{t I_\Omega} S_\omega \tag{3.27}$$

where τ includes the St. Venant torsional shear stress.

Hence, S_ω is obviously

$$S_\omega = S_{\Omega 0} - \frac{I_\Omega}{T_\Omega} q_w + \frac{I_\Omega}{T_\Omega} q_s \tag{3.28}$$

where q_s is the shear flow due to St. Venant torsion in the particular cell of the multicellular section.

3.10   Illustrative Examples

Example 3.1 In Fig. 3.7, C is the centroid and S is the shear center. An arbitrary pole P is located as shown, and the coordinates of S from pole P are X_{1S} and Y_{1S}. Show that over the contour shown in the sketch

$$\omega_s = \omega_P - y_{1S}(x - x_0) + x_{1S}(y - y_0)$$

Solution: We have using Eq. 3.5.

$$\omega_s = \omega_P + (y_P - y_S)(x - x_0) - (x_P - x_S)(y - y_0)$$
$$= \omega_P - y_{1S}(x - x_0) + x_{1S}(y - y_0)$$

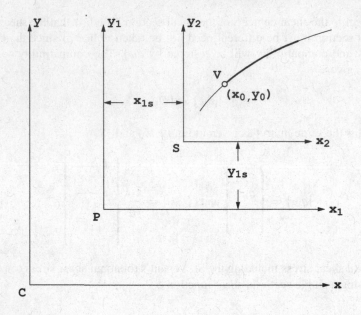

Fig. 3.7 Example 3.1

Example 3.2 The channel shown in Fig. 3.8a has a uniform wall thickness t. Plot the unit warping with the pole and sectorial origin located at the left bottom corner shown

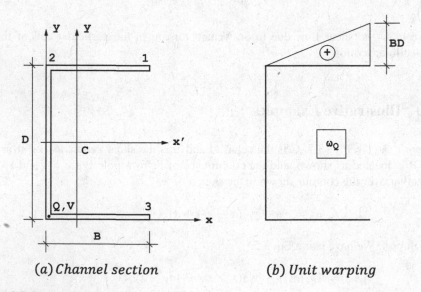

(a) *Channel section*   (b) *Unit warping*

Fig. 3.8 Example 3.2, 3.3

Solution: We have $\omega_V = 0$ since V is the initial point. The warping over Q3 and Q2 are zero since these plates pass through the pole and hence their tangential distance $h = 0$. Finally assuming clockwise sweep of sectorial areas as positive, we have

$$\omega_1 = \int_O^B D ds = BD$$

The unit warping is sketched in Fig. 3.8b

Example 3.3 Determine the normalized unit warping and hence the shear center of the thin-walled channel shown in Fig. 3.8.

Solution: The distributions of x- and y-coordinates of all points on the contour of the channel are shown in Fig. 3.9. Referring to Fig. 3.3 we have, using the product integrals of rectangle (*unit height*) and triangle in Eq. (3.7)

$$x_C = \frac{\int 1.x dA_s}{A_s} = \frac{[2]\frac{1}{2}(B)(1)(B)t}{(2B+D)t} = \frac{B^2}{2B+D}$$

Similarly using the product integrals

$$y_C = \frac{\left[\frac{1}{2}(D)(1)(D) + (B)(1)(D)\right]t}{(2B+D)t} = \frac{D}{2}$$

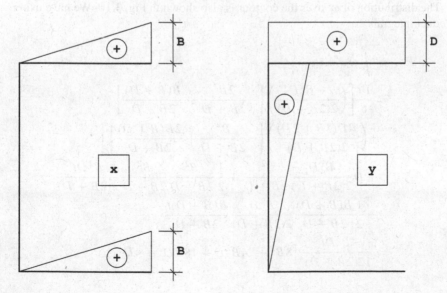

Fig. 3.9 Distribution of x, y

With these values of x_C and y_C, the distributions of x' and y' coordinates of all points on the contour can be generated as shown in Fig. 3.10. We now have

$$I_{x'} = \int y'y'\,\mathrm{d}A_s = 2\left(\frac{1}{3}\right)\left(\frac{D}{2}\right)\left(\frac{D}{2}\right)^2 t + 2(B)\left(\frac{D}{2}\right)^2 t$$

$$= \frac{D^2 t(D+6B)}{12}$$

$$I_{y'} = \int x'x'\,\mathrm{d}A_s = Dt\left[\frac{B^2}{2B+D}\right]^2 + 2\left(\frac{1}{3}\right)\frac{B^2 t}{2B+D}\left[\frac{B^2}{2B+D}\right]^2$$

$$+ 2\left(\frac{1}{3}\right)\frac{B(B+D)t}{2B+D}\left[\frac{B(B+D)}{2B+D}\right]^2$$

$$= \frac{B^3 t\left(4B^3 + 12B^2 D + 9BD^2 + 2D^3\right)}{3(2B+D)^3}$$

$$I_{x'y'} = 0$$

The normalized unit warping about the pole Q is

$$\omega' = \omega - \frac{\int \omega\,\mathrm{d}A_s}{\mathrm{d}A_s}$$

$$= \omega - \frac{\frac{1}{2}(B)(1)(BD)t}{((2B+D)t)} = \omega - \frac{B^2 D}{2(2B+D)}$$

The distribution of ω' over the contour is also shown in Fig. 3.10. We have using product integrals

$$F_{\omega'x'} = \int \omega'x'\,\mathrm{d}A_s$$

$$= \frac{Bt}{6}\left[\left(\frac{-B^2 D}{2(2B+D)}\right)\left\{-\frac{2B^2}{2B+D} + \frac{B(B+D)}{2B+D}\right\}\right.$$

$$\left. + \left(\frac{BD(3B+2D)}{2(2B+D)}\right)\left\{-\frac{B^2}{2B+D} + \frac{2B(B+D)}{2B+D}\right\}\right]$$

$$+ D\frac{B^2 D}{2(2B+D)}\frac{B^2 t}{2B+D} + \frac{1}{2}\frac{B^2}{2B+D}\frac{B^2}{2B+D}\frac{B^2 Dt}{2(2B+D)}$$

$$- \frac{1}{2}\frac{B(B+D)}{2B+D}\frac{B^2 D}{2(2B+D)}\frac{B(B+D)t}{2B+D}$$

$$= \frac{B^3 Dt}{12(2B+D)^3}\left[8B^3 + 24B^2 D + 18BD^2 + 4D^3\right]$$

$$F_{\omega'y'} = \int \omega'y'\,dA_s$$

$$= \frac{Bt}{6}\left[\frac{-B^2D}{2(2B+D)}\left\{D+\frac{D}{2}\right\} + \frac{BD(3B+2D)}{2(2B+D)}\left\{\frac{D}{2}+D\right\}\right] + B\frac{D}{2}\frac{B^2Dt}{2(2B+D)}$$

$$= \frac{B^2D^2t}{4}$$

The coordinates of the shear center can be found using Eqs. (3.15) and (3.16). Thus,

$$x'_s = -\frac{B^2}{2B+D} - \frac{B^2D^2t}{4}\frac{12}{D^2t(D+6B)}$$

$$= -\frac{4B^2(D+3B)}{(D+2B)(D+6B)}$$

$$y'_s = -\frac{D}{2} + \frac{B^3Dt\left(8B^3 + 24B^2D + 18BD^2 + 4D^3\right)}{12(2B+D)^3}\frac{3(2B+D)^3}{B^3t(4B^3 + 12B^2D + 9BD^2 + 2D^3)}$$

$$= 0$$

Example 3.4 Determine the principal normalized unit warping for the channel shown in Fig. 3.8, and hence determine the warping moment of inertia.

Solution: We have from Eq. 3.16

$$\omega_n = \omega' + \left(-\frac{D}{2} - 0\right)x' - \left(-\frac{B^2}{(2B+D)} + \frac{4B^2(D+3B)}{(D+2B)(D+6B)}\right)y'$$

$$= \omega' - \frac{D}{2}x' - \frac{3B^2}{(D+6B)}y'$$

The ω_n values can be calculated from the foregoing expression. As an example, for point 3, we have

$$\omega_{n3} = -\frac{B^2D}{2(2B+D)} - \frac{D}{2}\left[\frac{B(B+D)}{2B+D}\right] - \frac{3B^2}{D+6B}\left(-\frac{D}{2}\right) = -\frac{BD}{2}\frac{3B+D}{D+6B}$$

The distribution of ω_n is shown in Fig. 3.10. We have

$$I_\Omega = 2\left[\frac{Bt}{6}\left\{\frac{3B^2D}{2(D+6B)}\left[\frac{6B^2D}{2(D+6B)} - \frac{BD}{2}\frac{D+3B}{D+6B}\right]\right.\right.$$

$$+\left.\left(-\frac{BD}{2}\frac{D+3B}{D+6B}\right)\left(\frac{3B^2D}{2(D+6B)} - BD\frac{D+3B}{D+6B}\right)\right\}\right]$$

$$+\frac{Dt}{6}\left[\frac{3B^2D}{2(D+6B)}\left\{\frac{6B^2D}{2(D+6B)} - \frac{3B^2D}{2(D+6B)}\right\}\right.$$

$$+\left.\left(-\frac{3B^2D}{2(D+6B)}\right)\left\{\frac{3B^2D}{2(D+6B)} - \frac{3B^2D}{D+6B}\right\}\right]$$

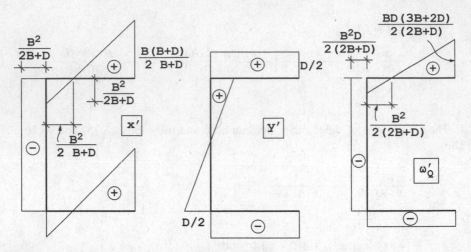

Fig. 3.10 Distribution of normalized unit warping (ω_n)

$$= \frac{B^3 D^2 t (2D + 3B)}{12(D + 6B)}$$

Example 3.5 For the channel shown in Fig. 3.8, determine the thickness warping moment of inertia and show that it is negligible compared to the contour warping moment of inertia derived in the previous problem.

Solution: we have the thickness warping given by Eq. (3.22)

$$\omega_t = -nq(s)$$

The distribution of $q(s)$ is shown in Fig. 3.11.
We have

$$I_{\Omega t} = \iint \omega_t^2 \, ds \, dt$$

$$= \int n^2 dt \int q^2 ds$$

$$= 2\left[\frac{1}{3}\frac{t}{2}\left(\frac{t}{2}\right)^2\right]\left[2\frac{B}{6}\left\{\frac{3B^2}{6B+D}\left(\frac{6B^2}{6B+D}+\frac{B(9B+D)}{6B+D}\right)\right.\right.$$

$$\left.\left.+\frac{B(9B+D)}{6B+D}\left(\frac{3B^2}{6B+D}+\frac{2B(9B+D)}{6B+D}\right)\right\}+2\left\{\frac{1}{3}\frac{D}{2}\left(\frac{D}{2}\right)^2\right\}\right]$$

$$= \frac{B^3 t^3}{18(6B+D)^2}\left(D^2 + 21BD + 117B^2\right) + \frac{D^3 t^3}{144}$$

We thus have

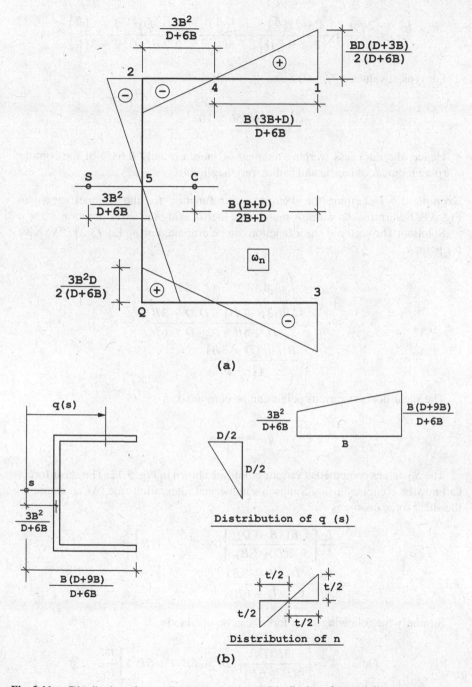

Fig. 3.11 a Distribution of normalized unit warping. **b** Distribution of normal distance

$$\frac{I_{\Omega t}}{I_\Omega} = \frac{2}{3}\left(\frac{t}{D}\right)^2 \frac{1 + 21\left(\frac{B}{D}\right) + 117\left(\frac{B}{D}\right)^2}{\{2 + 3\left(\frac{B}{D}\right)\}\{1 + 6\left(\frac{B}{D}\right)\}} + \frac{(t/D)^2}{12(B/D)^3}\frac{1 + 6\left(\frac{B}{D}\right)}{2 + 3\left(\frac{B}{D}\right)}$$

For typical values of $\frac{B}{D} = 0.5$ and $\frac{t}{D} = 0.05$

$$\frac{I_{\Omega t}}{I_\Omega} = 6.756 \times 10^{-3}$$

Hence, the thickness warping moment of inertia is only 0.68% of the contour warping moment of inertia and is thus very negligible.

Example 3.6 Determine the sectorial shear function for the channel shown in Fig. 3.8. Determine the warping moment of inertia using the shear function.

Solution: The sectorial shear function can be evaluated using Eq. (3.23). We have for point 4,

$$S_{\Omega 4} = \int \omega_n dA_s$$

$$= \frac{1}{2}\frac{B(3B + D)}{D + 6B}\frac{BD}{2}\frac{D + 3B}{D + 6B}t$$

$$= \frac{B^2 Dt}{4}\frac{(D + 3B)^2}{(D + 6B)^2}$$

The shear flows at various points can be computed as

$$q_w = \frac{T_\Omega}{I_\Omega}S_\Omega$$

The S_Ω values computed at various points are shown in Fig. 3.12. The shear forces can now be computed using Simpson's numerical integration rule. As an example, the shear force on 4–1 is

$$= \frac{T_\Omega}{I_\Omega}\left[\frac{1}{3}\frac{B(3B + D)t}{2(D + 6B)}\left\{0 + 4.\frac{3}{4}R + R\right\}\right]$$

$$= \frac{T_\Omega}{I_\Omega}\frac{B^3 Dt}{6}\frac{(D + 3B)^3}{(D + 6B)^3}$$

Similarly the following shear forces can be obtained

$$\text{On } 2 - 4 = \left[\frac{3B^4 Dt}{4(D + 6B)^3}\left(D^2 + 6BD + 6B^2\right)\right]\frac{T_\Omega}{I_\Omega}$$

$$\text{On } 5 - 2 = \left[0.024\frac{B^2 D^3 t}{(D + 6B)}\right]\frac{T_\Omega}{I_\Omega}$$

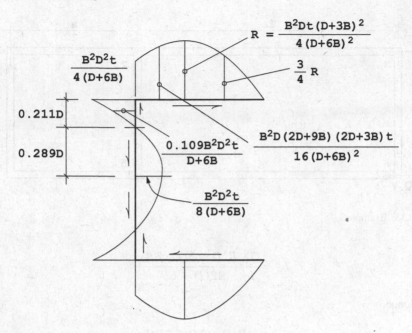

Fig. 3.12 Sectorial shear function values for the channel

$$(over \, 0.211D)$$

$$On \, 5 - 2 = \left[0.048 \frac{B^2 D^3 t}{(D + 6B)} \right] \frac{T_\Omega}{I_\Omega}$$

$$(over \, 0.289D)$$

The total shear force on 2–1 due to warping torsion is equal to (summing forces On $2 - 4$ and $4 - 1$)

$$\frac{B^3 Dt}{12(D + 6B)^3} \left[2(D + 3B)^3 + 9B\left(D^2 + 6BD + 6B^2\right) \right] \frac{T_\Omega}{I_\Omega}$$

$$= \frac{B^3 Dt}{12(D + 6B)^3} \left[(2D + 3B)(D + 6B)^2 \right] \frac{T_\Omega}{I_\Omega}$$

$$= \left[\frac{B^3 Dt (2D + 3B)}{12(D + 6B)} \right] \frac{T_\Omega}{I_\Omega}$$

The warping torsion can be obtained by taking the moments of all shear forces about the shear center. Thus

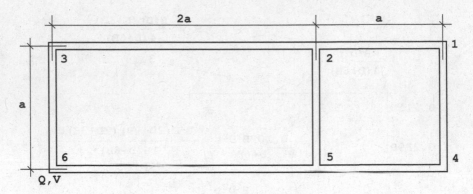

Fig. 3.13 Example 3.7

$$T_\Omega = \frac{T_\Omega}{I_\Omega} \frac{B^3 D^2 t(2D + 3B)}{12(D + 6B)}$$

Hence,

$$I_\Omega = \frac{B^3 D^2 t(2D + 3B)}{12(D + 6B)}$$

This checks with the value of the warping moment of inertia obtained in Example 3.4.

Example 3.7 Locate the shear center of the twin-celled cross section as shown in Fig. 3.13. Hence, determine the warping moment of inertia and the sectorial shear function.

Solution: The origin of the x–y system is chosen at point 6. We have the centroidal distance located as

$$x_c = \frac{2(3at)(1.5a) + at(2a) + at(3a)}{9at} = \frac{14a}{9}$$

Also,

$$I_{x'} = 3 \times \frac{1}{12} t a^3 + 2(3at)\left(\frac{a^2}{4}\right) = \frac{7a^3 t}{4}$$

$$I_{y'} = \frac{83a^3 t}{9}$$

The principal centroidal axes are themselves x' and y' since the section has an axis of symmetry.

To determine the unit warping about Q, we must know the modified shear flows due to St. Venant torsion viz., ψ_1 and ψ_2. They can be computed by solving the

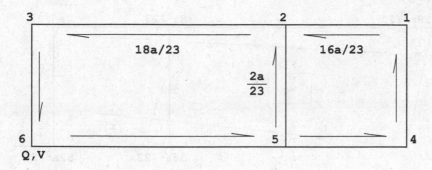

Fig. 3.14 Modified shear flow due to St. Venant Torsion

following equations

$$\frac{6a}{t}\psi_1 - \frac{a}{t}\psi_2 = 4a^2$$

$$-\frac{a}{t}\psi_1 + \frac{4a}{t}\psi_2 = 2a^2$$

Giving $\psi_1 = \frac{18at}{23}$ and $\psi_2 = \frac{16at}{23}$. These are shown in Fig. 3.14. Choosing the initial point, V (*point of zero warping*) as shown in Fig. 3.14, we have using Eq. 3.21.

$$\omega_6 = 0 \quad \text{(as assumed)}$$

$$\omega_3 = 0 + \left(0 - \frac{18a}{23}\right)a = -\frac{18a^2}{23} \quad \text{(path } 6 - 3)$$

In the foregoing h_{sc} is zero. The positive sign is used for h_{sc} when the movement from i to j (in this case 6–3) produces clockwise rotation about the pole Q. When the St. Venant shear flow opposes the movement direction, it is used with a negative sign as in the foregoing. When it assists the movement, a plus sign is used. Keeping this in mind, the unit warping at various points is computed as follows:

$$\omega_2 = -\frac{18a^2}{23} + \left(a - \frac{18a}{23}\right)2a = -\frac{8a^2}{23} \quad \text{(path } 3 - 2)$$

$$\omega_5 = 0 + \left(\frac{18a}{23}\right)2a = \frac{36a^2}{23} \quad \text{(path } 6 - 5)$$

$$\omega_2 = \frac{36a^2}{23} + \left(-2a + \frac{2a}{23}\right)a = -\frac{8a^2}{23} \quad \text{(path } 5 - 2)$$

(which checks)

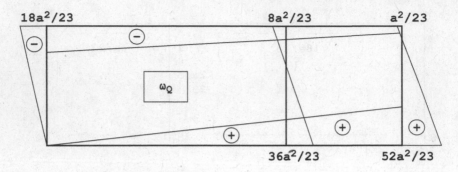

Fig. 3.15 Distribution of unit warping

$$\omega_4 = \frac{36a^2}{23} + \left(0 + \frac{16a}{23}\right)a = \frac{52a^2}{23} \quad (\text{path } 5-4)$$

$$\omega_1 = \frac{52a^2}{23} + \left(-3a + \frac{16a}{23}\right)a = -\frac{a^2}{23} \quad (\text{path } 4-1)$$

$$\omega_2 = -\frac{a^2}{23} + \left(-a + \frac{16a}{23}\right)a = -\frac{8a^2}{23} \quad (\text{path } 1-2)$$

(which checks)

The distribution of unit warping, ω_Q, is shown in Fig. 3.15.
The normalized unit warping about the pole Q can be obtained as

$$\omega'_Q = \omega_Q - \frac{\int \omega_Q dA_s}{A_s}$$

We have

$$\int \omega_Q dA_S = \frac{1}{2}(2a)\left(\frac{36a^2}{23}\right)t + \frac{at}{2}\left(\frac{88a^2}{23}\right) + \frac{at}{2}\left(\frac{51a^2}{23}\right) + \frac{at}{2}\left(\frac{28a^2}{23}\right)$$

$$-\frac{1}{2}a\left(\frac{18a^2}{23}\right)t - \frac{2at}{2}\left(\frac{26a^2}{23}\right) - \frac{at}{2}\left(\frac{9a^2}{23}\right) = \frac{80a^3t}{23}$$

Hence,

$$\omega'_Q = \omega_Q - \frac{80a^3t}{23(9at)} = \omega_Q - \frac{80a^2}{207}$$

The ω'_Q values computed from the foregoing are represented in Fig. 3.16. Using product integrals, we can obtain

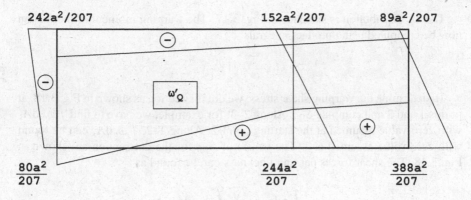

242a²/207   152a²/207   89a²/207

80a²/207   244a²/207   388a²/207

Fig. 3.16 ω'_Q values

$$F_{\omega'Y'} = -\frac{583.5}{207}a^4t$$

The shear center is thus located as

$$x'_S = -\frac{14a}{9} + \frac{583.5}{207}a^4t\frac{4}{7a^3t}$$

$$= \left(-\frac{14}{9} + \frac{2334}{1449}\right)a = 0.055a$$

The principal normalized unit warping values can now be obtained

$$\omega_n = \omega' - \frac{a}{2}x' + \frac{778}{483}ay'$$

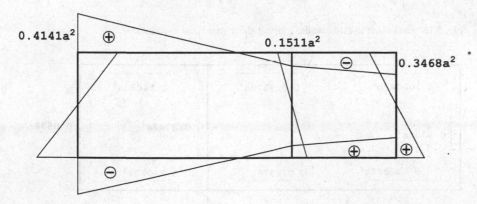

0.4141a²   0.1511a²   0.3468a²

Fig. 3.17 Distribution of ω_n

The ω_n distribution is sketched in Fig. 3.17. The warping moment of inertia can now be computed using product integrals

$$I_\Omega = 0.411a^5t$$

To determine the warping shear stress, we cut the section as shown in Fig. 3.18, at points 1 and 2 and compute $S_{\Omega 0}$. Along 2–3, for example, we have to find $\int \omega_n \, dA_s$ with zero value assumed at the starting point 2. Along 1–2, $\int \omega_n dA_s$ can be found with zero value assumed at 1. The values of $S_{\Omega 0}$ for the cut section are shown in Fig. 3.18. The shear forces per unit thickness can be found as

$$\int \tau \mathrm{d}s = \frac{T_\Omega}{t I_\Omega} \int S_{\Omega 0} \mathrm{d}s$$

These values are shown in Fig. 3.19, without the common multiplier T_Ω / I_Ω. The $\int \tau \mathrm{d}s$ values for cells 1 and 2 can be found. The following compatibility equations

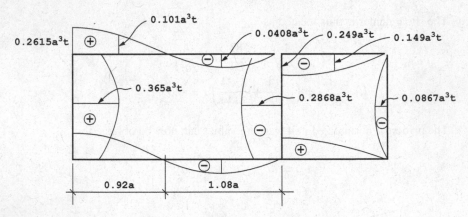

Fig. 3.18 Cuts made to calculate the warping shear stress and its values

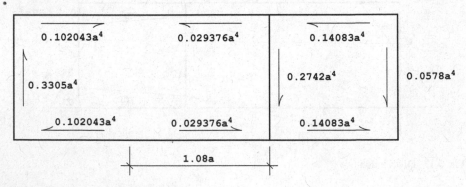

Fig. 3.19 Shear force per unit thickness

can be written for the cells, assuming the correcting warping shear flow as positive anticlockwise in each cell:

$$\frac{6a}{t}q_{W1} - \frac{a}{t}q_{w2} = \frac{T_\Omega a^4}{I_\Omega}(0.750034)$$

$$\frac{-a}{t}q_{w1} + \frac{4a}{t}q_{w2} = \frac{T_\Omega a^4}{I_\Omega}(-0.49806)$$

Solving

$$q_{w1} = 0.10879\frac{T_\Omega a^3 t}{I_\Omega}$$

$$q_{w2} = -0.09732\frac{T_\Omega a^3 t}{I_\Omega}$$

The actual warping shear stress at any point can be found as

$$\tau_W = \frac{T_\Omega}{tI_\Omega}S_{\Omega 0} + \frac{q_w}{t}$$

The warping shear stress distribution is sketched in Fig. 3.20, the common multiplying factor to this figure being $T_\Omega a^3/I_\Omega$.

The shear forces in the cut section and those due to correcting shear flows are shown in Fig. 3.21. The former forces can be obtained by multiplying the $\int \tau ds$

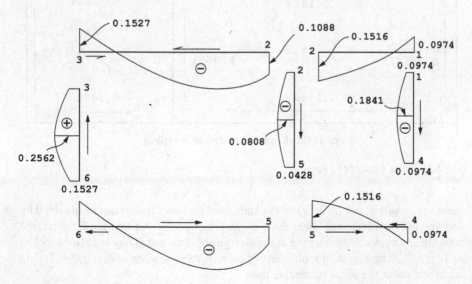

Fig. 3.20 Warping shear stress distribution

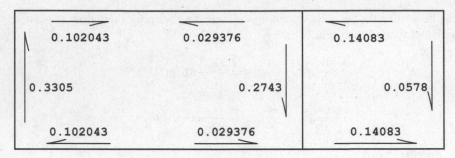

a. Shear forces in cut section

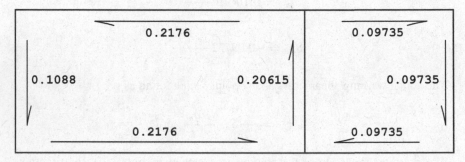

b. Shear forces due to correcting shear flow

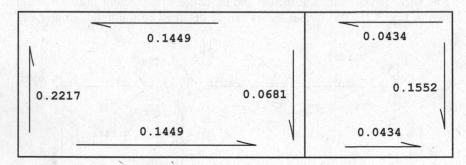

c. Total shear forces due to warping

Fig. 3.21 Shear force due to warping

values (i.e., values in Fig. 3.19) by thickness and the latter forces can be obtained by multiplying the correcting shear flows by length of plates. Adding these values, the total shear forces due to warping in the section are obtained. They are also sketched in Fig. 3.21. The common multiplier for each of these figures is $T_\Omega a^4 t / I_\Omega$. Taking moments about the shear center, we have

$$T_\Omega = \frac{T_\Omega a^5 t}{I_\Omega}[(0.0681)(0.39) + (0.1552)(1.39) + 0.2227(1.61) - 0.1883(1.0)]$$

$$= 0.411\frac{T_\Omega a^5 t}{I_\Omega}$$

This gives

$$I_\Omega = 0.411 a^5 t$$

This agrees with the value of warping moment of inertia computed earlier.

References

1. VZ Vlasov 1961 Thin Walled Elastic Beams Israel Program for Scientific Transaction Ltd Jerusalem
2. NW Murray 1986 Thin Walled Structures Clarendon Press Oxford
3. JJ Connor 1976 Analysis of Structural Member Systems The Ronald Press Company New York

Chapter 4
Theories of Torsion

4.1 Introduction

Straight, thin-walled, prismatic beams find extensive applications in engineering. Theories of thin-walled beams with open and closed sections are presented in this chapter. In these theories, the distortion of the cross section is neglected. This means that the shape of the cross section does not change, i.e., there is no deformation of the contour in the plane of the cross section. The entire cross section thus rotates about a point in its plane, and the angle of rotation is thus the same for all the plates forming the contour of the cross section and which is also equal to the angle of rotation of the entire cross section.

The crucial assumption given in foregoing assumes the cross section to be rigid in its own plane, i.e., there is no distortion. Deplanation of the cross section (i.e., displacements normal to the cross section), however, occurs, and these displacements are called warping. In order to realize the assumptions, it is necessary that the beam has been provided with diaphragms at regular intervals along its length. These so-called diaphragms are assumed to be perfectly rigid in their own plane but completely flexible out of their plane. Physically, these diaphragms are provided in the form of bulkheads, transverse frames or a network of rings and stringers. The thin-walled beam theories necessitate the prevention of distortion for them to apply, but they cannot give any information regarding the effect of the diaphragms, i.e., their spacing and rigidity. This question can be answered only by a finite element analysis of the 3D structure, discarding the beam assumption. Such an analysis could consider the distortion of the cross section and the spacing and rigidity of the diaphragms. The analysis would, however, be more costly and time consuming, and hence, the beam analysis developed in this chapter is more often preferred. Local effects of the diaphragms must, however, be accounted separately. It is, however, known that diaphragms are more important in closed than in open cross sections. Also the longer the beam and thicker the contour plates, the less important are the effects of the diaphragms.

K. Rajagopalan, *Torsion of Thin Walled Structures*,
https://doi.org/10.1007/978-981-16-7458-7_4

4.2 Basic Assumptions

Theories for open and closed profiles have much in common except for the cross-sectional properties. In open profile, Vlasov made the assumption that the shearing strain at the center of the contour is zero. The St. Venant shearing strain is actually zero at the center line (due to linear variations of shear stress across thickness). In addition, warping shearing strain is also assumed zero, so that the total shearing strain is zero. This means that warping shearing stresses cannot be determined using $\tau_W = G\gamma_W$ but must be obtained using equilibrium (The Bernoulli–Euler beam theory suffers from the same defect) [1].

For closed profile, the theory had been developed by von Karman and Christensen. They also neglected the warping shearing strain and assumed that the shearing strain at the centerline of the contour is due to the Bredt shear stress.

Some sections undergo large warping strains γ_W, and hence, their neglect in the theory would make the theory very approximate for such cases. In these cases, the Vlasov or von Karman approximate theories can be discarded and a more accurate theory which includes the warping shearing strain developed by Benscoter can be used.

4.3 Equilibrium Equations

A representation of the thin-walled beam is shown in Fig. 4.1. The only loading considered is a distributed torque m_z. Distributed surface loads are not considered to make the derivation less obscure. Their effects can be included without difficulty if necessary.

A small portion dz of the beam between stress-free ends is also shown in Fig. 4.1. Since the beam is twisted, St. Venant torque T_s would be present. The only stresses in the problem are the normal stress σ_z and the shear stress τ_{zs}. The force equations of equilibrium and the moment equation of equilibrium about an arbitrary point B are

$$\int_{A_s} \frac{\partial(\sigma_z t)}{\partial z} dz\, ds = 0$$

$$\int_{A_s} \frac{\partial(\tau_{zs} t)}{\partial z} \cos \psi\, dz\, ds = 0$$

$$\int_{A_s} \frac{\partial(\tau_{zs} t)}{\partial z} \sin \psi\, dz\, ds = 0$$

a

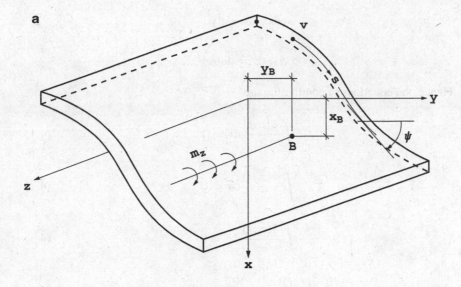

b

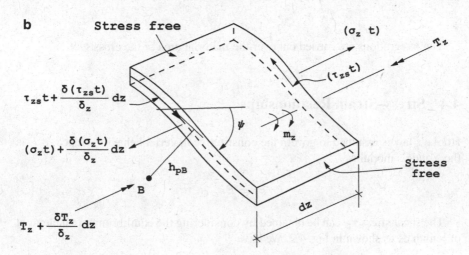

Fig. 4.1 a Representation of thin-walled beam. **b** Small portion of the beam

$$\int_{A_s} \frac{\partial(\tau_{zs}t)}{\partial z} h_{pB} dz\, ds + \frac{\partial T_s}{\partial z} dz + m_z dz = 0$$

We, however, have

$$\sin\psi\, ds = dx$$

$$\cos \psi \mathrm{d}s = \mathrm{d}y$$

$$h_{pB}\mathrm{d}s = \mathrm{d}\omega_B$$

Hence, the equations of equilibrium are

$$\int_{A_s} \frac{\partial [\sigma_z t]}{\partial z} \mathrm{d}s = 0 \tag{4.1}$$

$$\int_{A_s} \frac{\partial (\tau_{zs} t)}{\partial z} \mathrm{d}x = 0 \tag{4.2}$$

$$\int_{A_s} \frac{\partial (\tau_{zs} t)}{\partial z} \mathrm{d}y = 0 \tag{4.3}$$

$$\int_{A_s} \frac{\partial (\tau_{zs} t)}{\partial z} \mathrm{d}\omega_B + \frac{\partial T_s}{\partial z} + m_z = 0 \tag{4.4}$$

The integrations are carried out over the material area of the cross section, A_s.

4.4 Stress–Strain Relationships

Hooke's law is assumed to govern the constitutive material behavior. Hence, if E is the Young's modulus,

$$\sigma_z = E \epsilon_z \tag{4.5}$$

The shear stress τ_{zs} can be obtained by considering the equilibrium of an element of length ds as shown in Fig. 4.2. We have

$$\frac{\partial (\sigma_z t)}{\partial z} + \frac{\partial (\tau_{zs} t)}{\partial s} = 0 \tag{4.6}$$

If the shear stress is zero at an arbitrary point V as shown in Fig. 4.2, we have on integration

$$\tau_{zs} = \frac{1}{t} \left[-\int_0^{S_0} \frac{\partial (\sigma_z t)}{\partial z} \mathrm{d}s + G(z) \right] \tag{4.7}$$

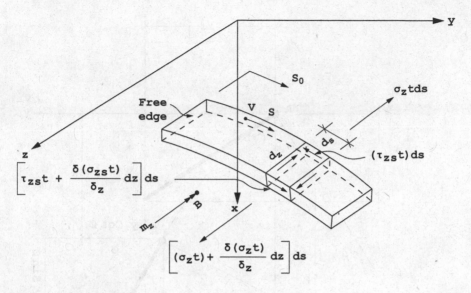

Fig. 4.2 Element of the beam

where $G(z)$ is a function of z only and independent of s. The shear stress can be expressed in terms of the longitudinal strain ϵ_z by using Eqs. (4.5) in (4.7).

4.5 Compatibility Equations

The cross section of the beam is bent and twisted under flexure and torsion. However, it is assumed that it retains its shape, and so all in-plane displacements can be expressed in terms of the cross-sectional rotation ϕ about a point.

The displacement ξ_i and η_i of a generic point i can be written with reference to Fig. 4.3. We have

$$\xi_i = \xi_A + y_i \sin\phi + x_i \cos\phi - x_i \tag{4.8}$$

$$\eta_i = \eta_A + y_i \cos\phi - x_i \sin\phi - y_i \tag{4.9}$$

When the angle of rotation ϕ is small, we can write

$$\xi_i = \xi_A + y_i\phi \tag{4.10}$$

$$\eta_i = \eta_A - x_i\phi \tag{4.11}$$

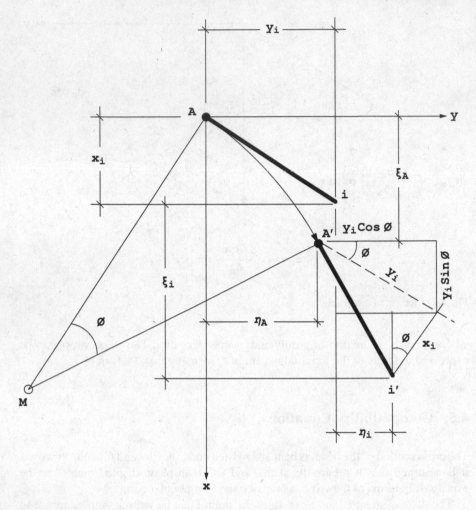

Fig. 4.3 Displacements

We can also express ξ_i and η_i in terms of displacement components of an arbitrary point B as shown in Fig. 4.4. We have

$$\xi_B = \xi_A + y_B \phi \tag{4.12}$$

$$\eta_B = \eta_A - x_B \phi \tag{4.13}$$

Thus,

$$\xi_i = \xi_B + (y_i - y_B)\phi \tag{4.14}$$

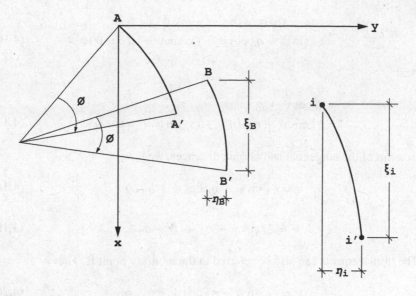

Fig. 4.4 Displacement components

$$\eta_i = \eta_B - (x_i - x_B)\phi \tag{4.15}$$

The displacements along the tangent and normal to the contour are denoted by v and u as shown in Fig. 4.5. It can be seen that

$$v = \eta_i \cos \psi + \xi_i \sin \psi$$

Fig. 4.5 Tangent and normal
displacement to contour

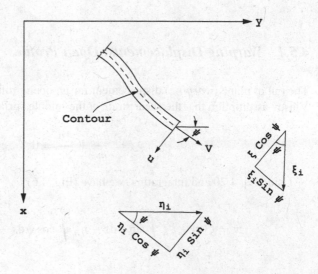

$$= (\eta_A - x_i\phi)\cos\psi + (\xi_A + y_i\phi)\sin\psi$$
$$= \xi_A\sin\psi + \eta_A\cos\psi + (y_i\sin\psi - x_i\cos\psi)\phi \qquad (4.16)$$

And

$$u = \xi_i\cos\psi - \eta_i\sin\psi$$
$$= \xi_A\cos\psi - \eta_A\sin\psi + (x_i\sin\psi + y_i\cos\psi)\phi \qquad (4.17)$$

In terms of the tangential and normal distances, we have

$$v = \xi_A\sin\psi + \eta_A\cos\psi + h_{pA}\phi \qquad (4.18)$$

$$u = \xi_A\cos\psi - \eta_A\sin\psi + q_A\phi \qquad (4.19)$$

The displacements can also be referred to the arbitrary point B. Thus,

$$v = \xi_B\sin\psi + \eta_B\cos\psi + h_{pB}\phi \qquad (4.20)$$

$$u = \xi_B\cos\psi - \eta_B\sin\psi + q_B\phi \qquad (4.21)$$

Where

$$h_{pB} = (y_i - y_B)\sin\psi - (x_i - x_B)\cos\psi \qquad (4.22)$$

$$q_B = (y_i - y_B)\cos\psi + (x_i - x_B)\sin\psi \qquad (4.23)$$

4.5.1   Warping Displacements: Open Profiles

The out of plane (*warping*) displacement for an open profile can be found from the Vlasov assumption that the shear strain at the middle surface is zero. Hence,

$$\gamma_{zs} = \frac{\partial w}{\partial s} + \frac{\partial v}{\partial z} = 0 \qquad (4.24)$$

Using Eq. 4.20 and integrating, we have (Fig. 4.6)

$$w = \zeta(z) - \xi_B'\int_0^s \sin\psi\,\mathrm{d}s - \eta_B'\int_0^s \cos\psi\,\mathrm{d}s - \phi'\int_0^s h_{pB}\,\mathrm{d}s$$

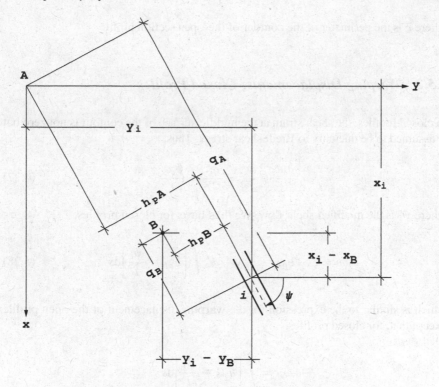

Fig. 4.6 Warping displacement open profile

$$= \zeta(z) - \xi'_B x - \eta'_B y - \phi' \omega_B \tag{4.25}$$

In the foregoing, $\zeta(z)$ is a constant of integration and represents a uniform longitudinal extension of the entire cross section (*due to axial forces on the bar if any*). The second and third terms represent bending, while the last term denotes torsion. The foregoing equation can be written in the matrix form as

$$w = \left\{ \zeta \ -\xi'_B \ -\eta'_B \ -\phi' \right\} \left\{ \begin{array}{c} 1 \\ x \\ y \\ \omega_B \end{array} \right\} \tag{4.26}$$

where

$$\left\{ \begin{array}{c} 1 \\ x \\ y \\ \omega_B \end{array} \right\} = \int_o^s \left\{ \begin{array}{c} \frac{1}{c} \\ \sin \psi \\ \cos \psi \\ h_{pB} \end{array} \right\} ds$$

where $\bar{c}$ is the perimeter of the contour of the open section profile.

4.5.2   Warping Displacements: Closed Profiles

In closed profiles, the shear strain at the middle surface of the contour is not zero but is assumed to be due only to Bredt shear stress. Thus,

$$\frac{\partial w}{\partial s} + \frac{\partial v}{\partial z} = \frac{q_i}{Gt} = \frac{\psi_i \phi'}{t} \tag{4.27}$$

where ψ_i is the modified shear flow. We thus have, for closed profiles,

$$w = \zeta - \xi_B' x - \eta_B' y - \phi' \int_0^s \left[h_{pB} - \frac{\psi_i}{t} \right] ds \tag{4.28}$$

which is similar to the expression for the warping displacement of the open profile except that, for closed profiles,

$$\omega_B = \int_0^s \left[h_{pB} - \frac{\psi_i}{t} \right] ds$$

The foregoing Eq. (4.28) can be written in the same matrix form Eq. (4.26) except for the fact that for closed profiles, ω_B is evaluated from the above expression.

4.6   Stress–Displacement Expressions

The strain in the longitudinal direction is simply

$$\begin{aligned} \epsilon_z &= \frac{\partial w}{\partial z} \\ &= \zeta' - \xi_B'' x - \eta_B'' y - \phi'' \omega_B \end{aligned} \tag{4.29}$$

The stress in the longitudinal direction is

$$\sigma_z = E \epsilon_z \tag{4.30}$$

The shear flow can be found from Eq. (4.7). Thus,

$$q = G(z) - \int\limits_{0}^{S_o} E\{t\zeta'' - t\xi_B''' x - t\eta_B''' y - t\phi''' \omega_B\}ds$$

$$= G(z) - E\zeta'' \int\limits_{0}^{S_o} tds + E\xi_B''' \int\limits_{0}^{S_o} txds$$

$$+ E\eta_B''' \int\limits_{0}^{S_o} tyds + E\phi''' \int\limits_{0}^{S_o} t\omega_B ds$$

$$= G(z) - E\zeta'' F + E\xi_B''' F_x + E\eta_B''' F_y + E\phi''' F_{\omega B} \tag{4.31}$$

In which, the section properties F, F_x, F_y, and $F_{\omega B}$ are defined in the previous chapter. For open profiles, at $s_o = 0$, $q = 0$, and hence, $G(z) = 0$. For closed profiles, a cut is made and compatibility at the cut is to be restored. The constants $G(z)$ would then be the modified shear flows in the cells to restore compatibility at the cuts. The methodology has already been demonstrated in the earlier chapter.

4.7 Differential Equations of Equilibrium

The equations of equilibrium can be expressed in terms of displacements $\xi, \eta, \zeta,$ and ϕ by integrating Eqs. (4.1) to (4.4) by parts. We get

$$\int \frac{\partial \sigma_z}{\partial Z} dA_s = 0 \tag{4.32}$$

$$\left[\frac{\partial \tau_{zs}}{\partial z} ty\right]_{\text{edge 1}}^{\text{edge 2}} - \int y \frac{\partial}{\partial s}\left[t\frac{\partial \tau_{zs}}{\partial z}\right] ds = 0 \tag{4.33}$$

$$\left[\frac{\partial \tau_{zs}}{\partial z} tx\right]_{\text{edge 1}}^{\text{edge 2}} - \int x \frac{\partial}{\partial s}\left[t\frac{\partial \tau_{zs}}{\partial z}\right] ds = 0 \tag{4.34}$$

$$\left[\frac{\partial \tau_{zs}}{\partial z} t\omega_B\right]_{\text{edge 1}}^{\text{edge 2}} - \int \omega_B \frac{\partial}{\partial s}\left[t\frac{\partial \tau_{zs}}{\partial z}\right] ds + \frac{\partial T_s}{\partial z} + m_z = 0 \tag{4.35}$$

In open profiles, the edges are free of stress, and so the first terms in the foregoing equations are zero. For closed profiles, the edges which are obtained by cuts carry equal stresses, and so these terms are also zero for closed profiles. Hence, using, Eqs. (4.30) into (4.32), we have

$$E\zeta'' \int dA_s - E\xi_B''' \int x dA_s - E\eta_B''' \int y dA_s - E\phi''' \int \omega_B dA_s = 0$$

which can be written as

$$E\zeta'' F - E\xi_B''' F_x - E\eta_B''' F_y - E\phi''' F_\omega = 0 \tag{4.36}$$

Similarly using Eq. (4.31) in Eqs. (4.33) to (4.35), we get

$$E\zeta''' F_y - E\xi_B'''' F_{xy} - E\eta_B'''' I_x - E\phi'''' F_{y\omega} = 0 \tag{4.37}$$

$$E\zeta''' F_x - E\xi_B'''' I_y - E\eta_B'''' F_{xy} - E\phi'''' F_{x\omega} = 0 \tag{4.38}$$

$$E\zeta''' F_\omega - E\xi_B'''' F_{x\omega} - E\eta_B'''' F_{y\omega} - E\phi'''' I_\omega + GJ\phi'' + m_z = 0 \tag{4.39}$$

The foregoing equation can be written in the matrix form

$$E\begin{bmatrix} F & F_x & F_y & F_\omega \\ F_y & I_y & F_{xy} & F_{x\omega} \\ F_y & F_{xy} & I_x & F_{y\omega} \\ F_\omega & F_{x\omega} & F_{y\omega} & I_\omega \end{bmatrix} \begin{Bmatrix} \zeta''' \\ -\xi_B'''' \\ -\eta_B'''' \\ -\phi'''' \end{Bmatrix} + G\begin{bmatrix} 0 & 0 & 0 & 0 \\ 0 & 0 & 0 & 0 \\ 0 & 0 & 0 & 0 \\ 0 & 0 & 0 & J \end{bmatrix} \begin{Bmatrix} 0 \\ 0 \\ 0 \\ -\phi'' \end{Bmatrix} = \begin{Bmatrix} 0 \\ 0 \\ 0 \\ -m_z \end{Bmatrix} \tag{4.40}$$

4.8 Decoupling

The decoupling of axial, bending, and torsional effects can be carried by diagonalizing the symmetric matrix in the first term of the foregoing expression. The decoupling is done in two stages. In the first stage, the axes x and y are shifted parallel to themselves with the origin of the new axes x' and y' being located at the centroid C as shown in Fig. 4.7. The starting point V is also shifted to make the last of the following integral zero.

In terms of the new axes, we have

$$F_{x'} = \int x' dA_s = 0$$

$$F_{y'} = \int y' dA_s = 0$$

$$F_{\omega_B'} = \int \omega' dA_s = 0$$

The matrix to be diagonalized now looks as

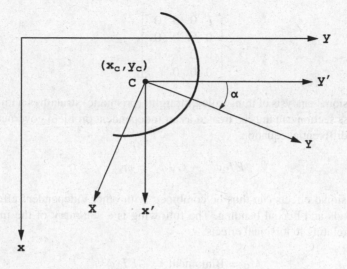

Fig. 4.7 Transformation for decoupling

$$\begin{bmatrix} F & 0 & 0 & 0 \\ 0 & I_{y'} & F_{x'y'} & F_{x'\omega'} \\ 0 & F_{x'y'} & I_{x'} & F_{y'\omega'} \\ 0 & F_{x'\omega'} & F_{y'\omega'} & I_{\omega'} \end{bmatrix}$$

In the second stage, the axes are rotated at the centroid to become the principal axes X and Y. This gives

$$F_{XY} = \int XY dA_s = 0$$

where the principal axes are located as

$$\tan 2\alpha = \frac{2F_{x'y'}}{I_{y'} - I_{x'}} \tag{4.41}$$

The pole is moved from B to the shear center. This makes

$$F_{X\omega_n} = \int X\omega_n dA_s = 0$$

$$F_{Y\omega_n} = \int Y\omega_n dA_s = 0$$

The diagonalized matrix is as follows

$$\begin{bmatrix} F & 0 & 0 & 0 \\ 0 & I_Y & 0 & 0 \\ 0 & 0 & I_X & 0 \\ 0 & 0 & 0 & I_\Omega \end{bmatrix} \tag{4.42}$$

The torsional analysis of thin-walled, straight, (prismatic) structures with open or closed cross section can thus be treated as an independent problem governed by the following differential equation.

$$E I_\Omega \phi'''' - G J \phi'' = m_z \tag{4.43}$$

The torsional effects can thus be combined with other independent effects due to axial loads and biaxial bending. The following is a summary of the important quantities relating to torsional effects:

$$M_\Omega = \text{Bimoment} = -E I_\Omega \phi'' \tag{4.44}$$

$$T_\Omega = \text{Warping torsion} = -E I_\Omega \phi''' \tag{4.45}$$

$$T_s = \text{St. Venant torsion} = G J \phi' \tag{4.46}$$

$$\text{Warping normal stress} = \sigma_w = \frac{M_\Omega}{I_\Omega} \omega_n \tag{4.47}$$

The bimoment is also given by

$$M_\Omega = \int_{A_s} \sigma_w \omega_n \mathrm{d}A_s \tag{4.48}$$

The warping normal stress is also given by

$$\sigma_w = -E \omega_n \phi'' \tag{4.49}$$

The warping shear stress is found as outlined in the previous chapter.

4.9 Benscoter's Theory

In the approximate theory, the warping strain is assumed zero, and hence, the shear stress cannot be obtained using the stress strain relationship, but must be obtained using an equilibrium condition. A more accurate theory by Benscoter removes this deficiency. The theory developed, however, gives similar differential equations except for a difference in cross-sectional properties involved in these equations.

In Benscoter's theory, the warping displacement is assumed to be proportional to a warping function $\theta(z)$ instead of $\phi'(z)$ as in the approximate theories. Thus,

$$w = -\theta(z)\omega_n(s) \tag{4.50}$$

The tangential displacement is given by the following due to the assumption that the cross section rotates as a rigid body

$$V = \phi(z)h_{sc}(s) \tag{4.51}$$

The longitudinal and shearing strains are

$$\epsilon_z = \frac{\partial \omega}{\partial z} = -\theta'\omega_n$$

$$\gamma_{zs} = \frac{\partial \omega}{\partial s} + \frac{\partial v}{\partial z} = h_{sc}\phi' - \theta\frac{\partial \omega_n}{\partial s}$$

However, we have

$$\omega_n = \int_o^s \left(h_{sc} - \frac{\psi_i}{t} \right) ds$$

Hence,

$$\frac{\partial \omega_n}{\partial s} = h_{sc} - \frac{\psi_i}{t}$$

Thus,

$$\gamma_{zs} = h_{sc}(\phi' - \theta) + \theta\frac{\psi_i}{t} \tag{4.52}$$

The stresses can be found from the stress–strain relationships

$$\sigma_z = E\epsilon_z = -E\theta'\omega_n \tag{4.53}$$

$$\tau_{zs} = G\gamma_{zs} = Gh_{sc}(\phi' - \theta) + G\theta\frac{\psi_i}{t} \tag{4.54}$$

We thus have

$$M_\Omega = \int\limits_{A_s} -E\theta'\omega_n.\omega_n \, dA_s = -EI_\Omega\theta' \tag{4.55}$$

and

$$T_Z = \int\limits_{A_s} h_{sc}(\tau_{zs}dA_s)$$

$$- \int Gh_{sc}^2(\phi' - \theta)d\Lambda_s + \int G\theta h_{sc} \cdot \frac{\psi_i}{t}dA_s$$

We define a new sectional property which may be called polar constant given by

$$I_h = \int h_{sc}^2 \, dA_s \tag{4.56}$$

Thus,

$$T_Z = GI_h(\phi' - \theta) + \int G\theta h_{sc}\psi_i ds$$

$$= GI_h(\phi' - \theta) + \sum_{i=1}^{n} G\theta\psi_i \int\limits_{i} h_{sc}ds$$

$$= GI_h(\phi' - \theta) + G\theta \sum 2A_i\psi_i$$

$$= GI_h(\phi' - \theta) + GJ\theta$$

Hence,

$$T_Z = GI_h\phi' - G\theta(I_h - J) \tag{4.57}$$

The equation of moment equilibrium about the longitudinal axis through the shear center gives

$$\frac{\partial T_Z}{\partial z} + m_Z = 0$$

Hence,

$$GI_h\phi'' - G\theta'(I_h - J) + m_Z = 0 \tag{4.58}$$

A second equilibrium equation can be obtained with a virtual displacement $\theta(z) = -1$. The virtual warping displacement is

$$W = \omega_n$$

The corresponding virtual shear strain is

$$\gamma = \frac{\partial w}{\partial s} = \frac{\partial \omega_n}{\partial s} = h_{sc} - \frac{\psi_i}{t}$$

The theorem of virtual work gives

$$\int \frac{\partial \sigma_z}{\partial z} \omega_n dA_s - \int \tau_{zs} \frac{\partial \omega_n}{\partial s} dA_s = 0$$

That is

$$-EI_\Omega \theta'' - \int \left[Gh_{sc}(\phi' - \theta) + G\theta \frac{\psi_i}{t} \right] \left[h_{sc} - \frac{\psi_i}{t} \right] dA_s = 0$$

which can be reduced to

$$-EI_\Omega \theta'' - G(I_h - J)(\phi' - \theta) = 0 \tag{4.59}$$

From Eq. (4.58),

$$\theta' = \frac{I_h}{I_h - J} \phi'' + \frac{m_Z}{G(I_h - J)} \tag{4.60}$$

Differentiating Eq. (4.59) with respect to z, we get

$$-EI_\Omega \theta''' - G(I_h - J)(\phi'' - \theta') = 0 \tag{4.61}$$

Adding the foregoing to Eq. (4.58), we have

$$-EI_\Omega \theta''' + GJ\phi'' + m_Z = 0 \tag{4.62}$$

Differentiating Eq. (4.60) twice with respect to z and using the result in Eq. (4.62), we have

$$-E\left[I_\Omega \frac{I_h}{I_h - J} \right] \phi'''' + GJ\phi'' = \frac{EI_\Omega}{G(I_h - J)} m_z'' - m_Z$$

which can be written as

$$-EI_{\Omega0} \phi'''' + GJ\phi'' = \frac{EI_{\Omega0}}{GI_h} m_z'' - m_Z \tag{4.63}$$

where

$$I_{\Omega0} = \frac{I_\Omega I_h}{I_h - J} \tag{4.64}$$

A comparison of Eqs. (4.63) with (4.43) shows that they are identical except for the value of I_Ω since for most of the distributed torsional loadings, m_z'' may be zero. Thus, in practical problems, the more accurate torsion theory can be simply obtained by replacing I_Ω by $I_{\Omega 0}$.

4.10 The Wagner Effect

A second-order term, known as the Wagner effect, must be added to the torsional moment term in Eq. (4.43) if it is significant. This torsional moment is produced by the normal stress on the warped (inclined) contour as shown in Fig. 4.8. The normal stress is inclined at an angle given by

$$r\phi' = \phi'\sqrt{(x' - x_s)^2 + (y' - y_s)^2} \tag{4.65}$$

Its contribution to the torsional moment about the shear center is

$$\overline{T} = \int \sigma r^2 \phi' \mathrm{d}A_s = \overline{K}\phi' \tag{4.66}$$

where the Wagner coefficient $\overline{K}$ is given as

Fig. 4.8 Warped contour (Wagner effect)

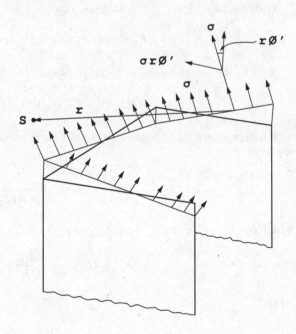

$$\overline{K} = \int \sigma r^2 \mathrm{d}A_s = \int \sigma \left\{ [x' - x_s]^2 + [y' - y_s]^2 \right\} \mathrm{d}A_s \qquad (4.67)$$

In terms of principal values, and taking N, M_X and M_Y as the axial force and the bending moments, the Wagner coefficient can be written as

$$\overline{K} = N \left\{ \frac{1}{A_s}(I_X + I_Y) + X_S^2 + Y_S^2 \right\} + M_X \left\{ \frac{1}{I_X} \left[\int YX^2 \mathrm{d}A_s + \int Y^3 \mathrm{d}A_s \right] - 2Y_S \right\}$$
$$+ M_Y \left\{ \frac{1}{I_Y} \left[\int X^3 \mathrm{d}A_s + \int XY^2 \mathrm{d}A_s \right] - 2X_S \right\}$$
$$+ \frac{M_\Omega}{I_\Omega} \left[\int \omega_n X^2 \mathrm{d}A_s + \int \omega_n Y^2 \mathrm{d}A_s \right] \qquad (4.68)$$

For a doubly symmetric cross section in which the shear center and the centroid coincide, we have

$$\overline{K} = \frac{N}{A_s}(I_X + I_Y) + \frac{M_\Omega}{I_\Omega} \left[\int \omega_n X^2 \mathrm{d}A_s + \int \omega_n Y^2 \mathrm{d}A_s \right] \qquad (4.69)$$

The normal stresses due to biaxial bending do not contribute to the Wagner effect in the case of doubly symmetric sections. The total torsional moment including the Wagner effect is thus $(GJ + \overline{K})\phi'$, and this must be used in place of $GJ\,\phi'$ in Eq. (4.43).

4.11 Review Questions

1. Why are diaphragms used in thin-walled beams? Are they necessary for theories described in this chapter to apply for the calculation of the warping stresses?
2. Recast the differential equations of Benscoter's theory into the following form

$$EI_\Omega \phi'''' - f_s GJ\phi'' = f_s m_z$$ and hence derive an expression for the warping shear parameter f_s.

4.12 Answers to Review Questions

1. Diaphragms are used at suitable intervals along the span to control the distortion of the cross section. As they do not prevent the deplanation (*warping*) of the beam cross section, they do not affect the warping stresses. However, they must be used for the theories of this chapter to apply because warping stresses calculated using these theories are based on the assumption that the cross section retains its shape and does not distort.

2. We have

$$\frac{E I_\Omega}{f_s} \phi'''' - GJ\phi'' = m_z$$

Comparing this with the Eq. (4.63) of Benscoter's theory, we have

$$I_{\Omega 0} = \frac{I_\Omega}{f_s}$$

Hence,

$$f_s = \frac{I_\Omega}{I_{\Omega 0}} = \frac{I_\Omega(I_h - J)}{I_\Omega I_h}$$

Thus,

$$f_s = 1 - \frac{J}{I_h}$$

Reference

1. Armenakas, A.E.: Advanced Mechanics of Materials and Applied Elasticity. CRC Press (2006)

Chapter 5
Analysis of Thin-Walled Structures

5.1 Introduction

The determination of stresses due to St. Venant and warping (*Vlasov*) torsions have
been dealt with earlier. This sectional analysis for stresses calls for the determination
of St. Venant and warping torsional moments at various sections of the structure due
to the loadings on the structure. This structural analysis problem toward the determination of the torsional moment distribution can be carried out by integrating the
differential equations developed earlier for simple cases and by numerical techniques
such as the finite element method for complex situations. This chapter deals with the
structural analysis involving differential equations.

5.2 St. Venant Torsion

When the warping rigidity of the structure is very small, the applied torque is carried
by St. Venant torsion. Typically, such a situation may exist when the value of μ is
very large where [1]

$$\mu = \frac{L}{d} \tag{5.1}$$

in which L is the representative span length and d is a parameter which had the
dimension of length given by

$$d = \sqrt{\frac{EI_\Omega}{GJ}} \tag{5.2}$$

The parameter d is known as the characteristic length and is often tabulated in
handbooks for typical structural steel sections.

© The Author(s), under exclusive license to Springer Nature Singapore Pte Ltd. 2022
K. Rajagopalan, *Torsion of Thin Walled Structures*,
https://doi.org/10.1007/978-981-16-7458-7_5

The differential equation for St. Venant torsion is simple and expresses the fact that the rate of change of rotation is directly proportional to the torsional moment at the section. It also can be extracted from Eq. (4.43)

$$GJ\phi'' = -m_z \tag{5.3}$$

The determination of the torsional moments at various sections can be obtained by integrating this differential equation. Two examples are considered in the following to illustrate the procedure.

A torsional member with fixed end conditions is shown in Fig. 5.1. It is subjected to a concentrated torque at the midsection. Since the distributed torque is zero, we have

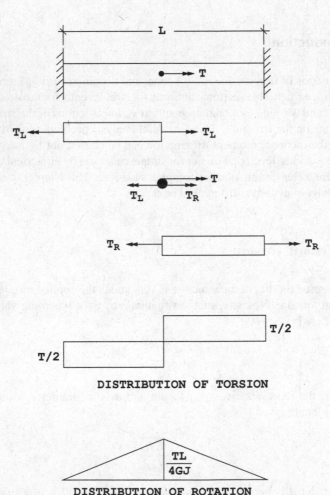

DISTRIBUTION OF TORSION

DISTRIBUTION OF ROTATION

Fig. 5.1 Torsion of a fixed member

$$0 < z < L/2 \quad GJ\phi'' = 0$$
$$L/2 < z < L \quad GJ\phi'' = 0$$

Integrating twice, we have

$$\left.\begin{array}{l} GJ\phi' = A \\ GJ\phi = Az + B \end{array}\right\} \text{ for } 0 < z < \frac{L}{2} \qquad \left.\begin{array}{l} GJ\phi' = C \\ GJ\phi = Cz + D \end{array}\right\} \text{ for } \frac{L}{2} < z < L$$

Since the ends are fixed, we have

$$\phi = 0 \text{ at } z = 0 \text{ Hence } B = 0$$
$$\phi = 0 \text{ at } z = L \text{ Thus } D = -CL$$

The rotation must be unique at the midsection. Hence,

$$A\frac{L}{2} = C\frac{L}{2} - CL \text{ Hence } A = -C.$$

Finally considering the equilibrium of the midsection point

$$T + T_R = T_L$$
$$T - A = A \quad \text{Hence } A = \frac{T}{2}$$

The complete solution of the problem is thus

$$\phi' = \frac{T}{2GJ} \text{ for } 0 < z < \frac{L}{2}$$
$$= -\frac{T}{2GJ} \text{ for } \frac{L}{2} < z < L \tag{5.4}$$

And

$$\phi = \frac{Tz}{2GJ} \text{ for } 0 < z < \frac{L}{2}$$
$$= \frac{T}{2GJ}(L - z) \text{ for } \frac{L}{2} < z < L \tag{5.5}$$

The distribution of torsional moments and rotations are sketched in Fig. 5.1.

The next example shown in Fig. 5.2 is a cantilever fixed at the left end and free at the other. The torque loading consists of a uniformly distributed torque t over the left half and $-t$ over the right half. We have for the structure

$$0 < z < L/2 \quad GJ\phi'' = -t$$
$$\frac{L}{2} < z < L \quad GJ\phi'' = t$$

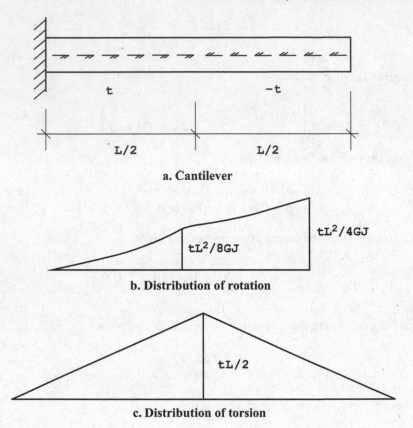

a. Cantilever

b. Distribution of rotation

c. Distribution of torsion

Fig. 5.2 Torsion of cantilever

Integrating we get

$$GJ\phi' = -tz + A \qquad GJ\phi' = tz + C$$
$$GJ\phi = -\frac{tz^2}{2} + Az + B \qquad GJ\phi = \frac{tz^2}{2} + Cz + D$$

At left end, $\phi = 0$. Hence, $B = 0$. At free end, the (sectional) torsion is zero. The torsion is also zero at the fixed end. Thus, at $z = 0$, $\phi' = 0$ Hence $A = 0$.

At $z = L$, $\phi' = 0$ Hence, $C = -tL$

The torsional rotation must be unique at the midsection. Hence,

$$-\frac{tL^2}{8} = \frac{tL^2}{8} - \frac{tL^2}{2} + D$$

Hence,

$$D = \frac{tL^2}{4}$$

The solution for the rotation is thus

$$\phi = -\frac{tz^2}{2GJ} \text{ for } 0 < z < \frac{L}{2}$$

$$= \frac{t(L^2 + 2z^2)}{4GJ} - \frac{tLz}{GJ} \quad \text{for } \frac{L}{2} < z < L \qquad (5.6)$$

The torsional moment is given by

$$T_s = tz \quad \text{for } 0 < z < \frac{L}{2}$$

$$= t(L-z) \quad \text{for } L/2 < z < L$$

The distributions of rotation and torsional moments are sketched in Fig. 5.3.

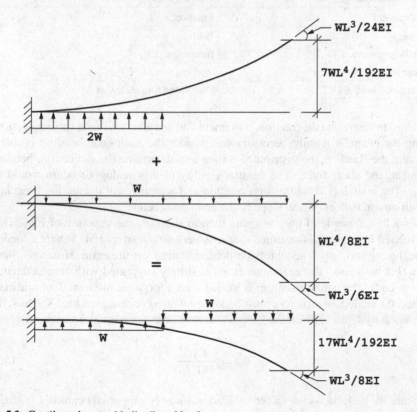

Fig. 5.3 Cantilever beam with distributed loads

5.3   VLASOV (*Warping*) Torsion

The warping torsion predominates over the entire structure when the value of μ is very small. The differential equation for pure warping torsion follows from Eq. (4.43)

$$EI_\Omega \phi'''' = m_z \tag{5.7}$$

An examination of Eq. (5.7) shows that there is a complete analogy between the warping torsion problem and a thin beam bending problem which is governed by the differential equation

$$EIV'''' = q$$

where V is the deflection and q is the load per unit length. The flexural rigidity is EI. The following are the analogous quantities.

Thin beam bending	Warping torsion
Deflection, V	Rotation, ϕ
Slope, V'	Warping, ϕ'
Bending moment, EIV''	Bimoment, $EI_\Omega \phi''$
Shear force, EIV'''	Warping torsion, $EI_\Omega \phi'''$
Distributed load, EIV''''	Distributed torque, $EI_\Omega \phi''''$

Thus, to determine the rotation, bimoment and warping torsion at various sections along the beam, it is only necessary to consider the analogous bending problem wherein the 'load' is the distributed torque and determine the deflection, bending moment, and shear force. The flexural rigidity of this analogous beam would be EI_Ω. The boundary conditions on rotation and warping will decide the boundary conditions on deflection and slope in the analogous beam.

As a first example of pure warping torsion analysis, the structure of Fig. 5.1 is considered in which it is assumed that μ is very small, so that St. Venant's torsion is negligible and only Vlasov torsion predominates over the entire structure. Since $\phi = 0$ at both ends, the analogous beam is simply supported with zero deflection at both ends. The actual structure is loaded with a torsional moment T at midspan. Hence, the analogous structure must be loaded with a concentrated load T. Thus, the rotation at midspan is given by

$$\phi_m = \frac{TL^3}{48EI_\Omega} \tag{5.8}$$

Since the deflection under a central load in a simply supported beam $WL^3/48EI$. The next example is the structure shown in Fig. 5.2. Here, both the rotation and warping are restrained at the left end. Hence, the analogous beam will have both deflection and slope zero at the left end, i.e., a fixed boundary condition. It can also be seen that the right end of the analogous beam will be free. The loading consists of a 'distributed load' of t over one-half and a 'distributed load' of $-t$ over the other half. The deflection at the free end under this loading is obtained as [Fig. 5.3].

$$\delta = \frac{17}{192} \frac{tL^4}{EI} \tag{5.9}$$

Hence, the torsional rotation at the free end

$$\phi_M = \frac{17}{192} \frac{tL^4}{EI_\Omega} \tag{5.10}$$

As a third example, consider the structure of Fig. 5.2 but with the addition that warping is restrained at the right end. The torsional rotation is, however, free to occur at this end. For these boundary conditions on the actual structure, the analogous beam will have a translation-free, rotation-fixed boundary condition at the right end. Under these boundary conditions, the deflection at the free end is given by

$$\delta = \frac{5}{192} \frac{tL^4}{EI} \tag{5.11}$$

Hence, the torsional rotation at the free end is

$$\phi_m = \frac{5}{192} \frac{tL^4}{EI_\Omega} \tag{5.12}$$

It is clear from the foregoing three examples that the analysis of structures under pure warping torsion can be easily carried out with the use of standard tables on beam deflections or analyzing easily the analogous beam using well-known methods of beam analysis.

5.4 Mixed Torsion

Mixed torsion problems can be analyzed using the differential Eq. (4.43)

$$EI_\Omega \phi'''' - GJ\phi'' = m_z \tag{5.13}$$

The solution with larger values of μ will coincide with the solution for St. Venant torsion (*except near warping restraints where it will be governed by warping torsion*), and the solution for mixed torsion with smaller values of μ will coincide with the solution of pure warping torsion.

As a first example, the structure of Fig. 5.1 is considered. The solution of the differential equation is

$$\phi = C_1 + C_2\left(\frac{z}{L}\right) + C_3 \sinh \frac{\mu z}{L} + C_4 \cosh \frac{\mu z}{L} \quad \text{for } 0 < z < \frac{L}{2} \tag{5.14}$$

$$\phi = C_5 + C_6\left(\frac{z}{L}\right) + C_7 \sinh \frac{\mu z}{L} + C_8 \cosh \frac{\mu z}{L} \quad \text{for } \frac{L}{2} < z < L \tag{5.15}$$

We have at the left end,

$$z = 0, \quad \phi = 0; \quad \text{from Eq. (5.15) we have } C_1 + C_4 = 0$$
$$z = 0, \quad \phi'' = 0; \quad \text{from Eq. (5.15) we have } C_4 = 0$$

Thus, $C_1 = C_4 = 0$
At the right end

$$z = L, \quad \phi = 0; \quad \text{from Eq. (5.16), } C_5 + C_6 + C_7 \sinh \mu + C_8 \cosh \mu = 0$$
$$z = L, \quad \phi'' = 0; \quad \text{from Eq. (5.16), } C_8 = -C_7 \tanh \mu$$

Thus,

$$C_5 = -C_6$$

The solution with the foregoing evaluation of some constants reduces to

$$\phi = C_2\left(\frac{z}{L}\right) + C_3 \sinh\left(\frac{\mu z}{L}\right) \text{ for } 0 < z < \frac{L}{2} \tag{5.16}$$

$$\phi = C_5\left(1 - \frac{z}{L}\right) + C_7\left[\sinh \frac{\mu z}{L} - \tanh \mu \cosh \frac{\mu z}{L}\right] \text{ for } \frac{L}{2} < z < L \tag{5.17}$$

At $z = L/2$, we have the equilibrium condition

$$T = T_L - T_R$$

Since

$$\text{torsion} = EI_\Omega \phi''' - GJ\phi'$$

We have

$$T = EI_\Omega\left[\frac{C_3\mu^3}{L^3} \cosh \frac{\mu}{2}\right] - GJ\left[\frac{C_2}{L} + \frac{C_3\mu}{L} \cosh \frac{\mu}{2}\right]$$

$$- EI_\Omega \left[\frac{C_7 \mu^3}{L^3} \cosh \frac{\mu}{2} - \frac{C_7 \mu^3 \tanh \mu}{L^3} \sinh \frac{\mu}{2} \right]$$

$$+ GJ \left[-\frac{C_5}{L} + \frac{C_7 \mu}{L} \cosh \frac{\mu}{2} - \frac{C_7 \mu \tanh \mu}{L} \sinh \frac{\mu}{2} \right]$$

However,

$$\frac{EI_\Omega}{GJ} = \frac{L^2}{\mu^2}$$

Hence, the foregoing gives

$$T = -\left(\frac{C_2 + C_5}{L} \right) GJ \tag{5.18}$$

For the continuity of ϕ at $z = \frac{L}{2}$, we have

$$\frac{C_2}{2} + C_3 \sinh \frac{\mu}{2} = \frac{C_5}{2} + C_7 \left[\sinh \frac{\mu}{2} - \tanh \mu \cosh \frac{\mu}{2} \right] \tag{5.19}$$

For the continuity of ϕ' at $z = \frac{L}{2}$, we get

$$\frac{C_2}{L} + \frac{C_3 \mu}{L} \cosh \frac{\mu}{2} = -\frac{C_5}{2} + C_7 \left[\frac{\mu}{L} \cosh \frac{\mu}{2} - \frac{\mu \tan h \mu}{L} \sinh \frac{\mu}{2} \right] \tag{5.20}$$

For the continuity of ϕ'' at $z = \frac{L}{2}$, we get

$$C_3 = C_7 \left[1 - \frac{\tanh \mu}{\tanh \frac{\mu}{2}} \right] \tag{5.21}$$

From Eqs. (5.19) and (5.20), we get

$$C_2 = C_5$$

Hence, from eqn. (5.18), we get

$$C_2 = C_5 = \frac{-TL}{2GJ} \tag{5.22}$$

Using Eqs. (5.22) and (5.21) in Eq. (5.20), we have

$$C_7 = -\frac{TL}{GJ} \frac{\sinh \frac{\mu}{2}}{\mu \tanh \mu} \tag{5.23}$$

All the constants in Eqs. (5.16) and (5.17) are thus known. These represent the solution of the mixed torsion problem of Fig. 5.1.

The next example considered is the structure of Fig. 5.2 with fixed boundary conditions for rotation and warping at the left end. The right end is free to warp and rotate. The solution of the differential equation for the problem is for the region $0 < z < \frac{L}{2}$

$$\phi = C_1 + C_2\left(\frac{z}{L}\right) + C_3 \sinh\left(\frac{\mu z}{L}\right) + C_4 \cosh\left(\frac{\mu z}{L}\right) - \frac{tz^2}{2GJ}$$

$$\phi' = \frac{C_2}{L} + \frac{C_3\mu}{L}\cosh\left(\frac{\mu z}{L}\right) + \frac{C_4\mu}{L}\sinh\left(\frac{\mu z}{L}\right) - \frac{tz}{GJ}$$

$$\phi'' = \frac{C_3\mu^2}{L^2}\sinh\left(\frac{\mu z}{L}\right) + C_4\frac{\mu^2}{L^2}\cosh\left(\frac{\mu z}{L}\right) - \frac{t}{GJ}$$

$$\phi''' = \frac{C_3\mu^3}{L^3}\cosh\left(\frac{\mu z}{L}\right) + \frac{C_4\mu^3}{L^3}\sinh\left(\frac{\mu z}{L}\right)$$

For the region $\frac{L}{2} < z < L$

$$\phi = C_5 + C_6\left(\frac{z}{L}\right) + C_7 \sinh\left(\frac{\mu z}{L}\right) + C_8 \cosh\left(\frac{\mu z}{L}\right) + \frac{tz^2}{2GJ}$$

$$\phi' = \frac{C_6}{L} + \frac{C_7\mu}{L}\cosh\left(\frac{\mu z}{L}\right) + \frac{C_8\mu}{L}\sinh\left(\frac{\mu z}{L}\right) + \frac{tz}{GJ}$$

$$\phi'' = \frac{C_7\mu^2}{L^2}\sinh\left(\frac{\mu z}{L}\right) + C_8\frac{\mu^2}{L^2}\cosh\left(\frac{\mu z}{L}\right) + \frac{t}{GJ}$$

$$\phi''' = \frac{C_7\mu^3}{L^3}\cosh\left(\frac{\mu z}{L}\right) + \frac{C_8\mu^3}{L^3}\sinh\left(\frac{\mu z}{L}\right)$$

Because of the nature of the loading, the torsion is zero at the left end. Hence, at $z = 0$

$$EI_\Omega \frac{C_3\mu^3}{L^3} - GJ\left[\frac{C_2}{L} + \frac{C_3\mu}{L}\right] = 0$$

Since

$$\frac{EI_\Omega}{GJ} = \frac{L^2}{\mu^2}$$

The foregoing gives $C_2 = 0$.
Also we have at the left end

$$z = 0, \quad \phi = 0, \quad \text{Hence } C_1 + C_4 = 0$$
$$z = 0, \quad \phi' = 0, \quad \text{Thus } C_2 = -C_3\mu$$

Hence, we have

$$C_1 = -C_4$$
$$C_2 = C_3 = 0$$

At the right end, we have zero bimoment, i.e., $z = L$, $\phi'' = 0$ which gives

$$C_7 \sinh \mu + C_8 \cosh \mu = -\frac{tL^2}{\mu^2 GJ} \qquad (5.24)$$

Also the torsional moment is zero at the right end; thus, at $z = L$,

$$EI_\Omega \left[\frac{C_7 \mu^3}{L^3} \cosh \mu + \frac{C_8 \mu^3}{L^3} \sinh \mu \right]$$
$$- GJ \left[\frac{C_6}{L} + \frac{C_7 \mu}{L} \cosh \mu + \frac{C_8 \mu}{L} \sinh \mu + \frac{tL}{GJ} \right] = 0 \qquad (5.25)$$

which gives

$$C_6 = -\frac{tL^2}{GJ}$$

For the continuity of ϕ, at $z = \frac{L}{2}$, we have

$$C_1 + C_4 \cosh \frac{\mu}{2} = C_5 + \frac{C_6}{2} + C_7 \sinh \frac{\mu}{2} + C_8 \cosh \frac{\mu}{2} + \frac{tL^2}{4GJ}$$

which gives

$$C_1 \left(1 - \cosh \frac{\mu}{2} \right) = C_5 + C_7 \sinh \frac{\mu}{2} + C_8 \cosh \frac{\mu}{2} \frac{tL^2}{4GJ} \qquad (5.26)$$

For the continuity of ϕ', at $z = \frac{L}{2}$, we have

$$\frac{C_4 \mu}{L} \sinh \frac{\mu}{2} - \frac{tL}{2GJ} = \frac{C_6}{L} + \frac{C_7 \mu}{L} \cosh \frac{\mu}{2} + \frac{C_8 \mu}{L} \sinh \frac{\mu}{2} + \frac{tL}{2GJ}$$

which gives

$$C_7 \cosh \frac{\mu}{2} + C_8 \sinh \frac{\mu}{2} = C_1 \sinh \frac{\mu}{2} \qquad (5.27)$$

Finally for the continuity of ϕ'', at $z = \frac{L}{2}$

$$C_7 \sinh \frac{\mu}{2} + C_8 \cosh \frac{\mu}{2} = -\frac{2tL^2}{\mu^2 GJ} + C_4 \cosh \frac{\mu}{2} \tag{5.28}$$

Solving the foregoing equations, we get

$$C_5 - C_1 = \frac{tL^2}{GJ}\left(\frac{1}{4} + \frac{2}{\mu^2}\right) \tag{5.29}$$

$$C_7 = \frac{2tL^2 \sinh \frac{\mu}{2}}{GJ\mu^2}$$

$$C_8 = -\frac{tL^2}{GJ\mu^2 \cosh \mu}\left[1 + 2 \sinh \mu \sinh \frac{\mu}{2}\right]$$

And

$$C_1 = \frac{tL^2}{GJ\mu^2}\left[-2 \cosh \frac{\mu}{2} + \frac{1}{\cosh \mu} + 2 \tanh \mu \sinh \frac{\mu}{2}\right]$$

The solution is complete since all the constants are determined.

As a third example, the structure of Fig. 5.2 is considered, but the right end is restrained against warping. The basic solutions for the two regions are the same as before, but the constants C_1 through C_8 are determined in the following way.

$\phi = 0$ at $z = 0$ This gives $C_1 + C_4 = 0$

$\phi' = 0$ at $z = 0$ Hence, $C_2 = -C_3\mu$

$\phi' = 0$ at $z = L$ by which we get

$$C_7 \cosh \mu + C_8 \sinh \mu = -\frac{C_6}{\mu} - \frac{tL^2}{\mu GJ}$$

$$\left[EI_\Omega \phi''' - GJ\phi'\right]_{z=0} = 0 \text{ which gives } C_2 = C_3 = 0$$

$$\left[EI_\Omega \phi''' - GJ\phi'\right]_{z=L} = 0 \text{ which shows } C_6 = -\frac{tL^2}{GJ}$$

For the continuity of ϕ, ϕ' and ϕ'' at $z = \frac{L}{2}$, we get

$$C_1\left(1 - \cosh \frac{\mu}{2}\right) = C_5 + C_7 \sinh \frac{\mu}{2} + C_8 \cosh \frac{\mu}{2} - \frac{tL^2}{4GJ}$$

$$C_7 \cosh \frac{\mu}{2} + C_8 \sinh \frac{\mu}{2} = -C_1 \sinh \frac{\mu}{2}$$

$$C_7 \sinh \frac{\mu}{2} + C_8 \cosh \frac{\mu}{2} = \frac{-2tL^2}{\mu^2 GJ} + C_4 \cosh \frac{\mu}{2}$$

Solving the foregoing equations, we get

$$C_1 - C_5 = -\frac{tL^2}{4GJ}\left[1 + \frac{8}{\mu^2}\right]$$

$$C_7 = \frac{2tL^2}{\mu^2 GJ}\sinh\frac{\mu}{2}$$

$$C_8 = -\frac{2tL^2}{\mu^2 GJ}\frac{\sinh\frac{\mu}{2}}{\tanh\mu}$$

$$C_1 = \frac{2tL^2}{\mu^2 GJ}\left[\frac{\sinh\frac{\mu}{2}}{\tanh\mu} - \sinh\frac{\mu}{2}\tanh\frac{\mu}{2} - \frac{1}{\cosh\frac{\mu}{2}}\right] \qquad (5.30a)$$

This completes the solution of the structure.

5.5 Warping Restraints

Restraints to warping occur at changes of cross section along the structure. The warping deformations at the junction of two dissimilar sections is a nonlinear function of the s-coordinate. The usual assumption of a linear function means that compatibility of warping deformations at the junction will not be satisfied. While higher-order warping functions can be used to develop the theories, the development becomes complex. Hence, the usual linear warping functions can be retained, but at the junctions compatibility is satisfied in an average or 'overall' sense through the use of warping compatibility factors. For example if 1 and 2 denotes the sections at the left and right side of a junction, we can write at the junction.

$$\phi_1 = \phi_2$$
$$\phi_1' = \gamma\phi_2'$$
$$\phi_1'' = \gamma\phi_2'' \qquad (5.31)$$

where γ is the warping compatibility factor for the junction. To find an expression for the warping compatibility factor, we equate to zero the integral of the square of the warping displacement errors at the junction. Thus,

$$\int\left[\omega_1\phi_1' - \omega_2\phi_2'\right]^2 ds = 0$$

which gives

$$\int\omega_1^2\phi_1'^2 ds + \int\omega_2^2\phi_2'^2 ds - 2\int\omega_1\omega_2\,\phi_1'\phi_2'\,ds = 0$$

The following constants A, B, and C can be defined

$$\int \omega_1^2 \phi_1'^2 \, ds = \phi_1'^2 \int \omega_1^2 ds = A \phi_1'^2$$

$$\int \omega_2^2 \phi_2'^2 \, ds = \phi_2'^2 \int \omega_2^2 \, ds = B \, \phi_2'^2$$

$$\int \omega_1 \omega_2 \phi_1' \phi_2' \, ds = \phi_1' \phi_2' \int \omega_1 \omega_2 \, ds = C \phi_1' \phi_2'$$

Now, we have

$$A \phi_1'^2 - 2C \phi_1' \phi_2' + B \, \phi_2'^2 = 0$$

From which ϕ_1' can be solved in terms of ϕ_2'. Thus,

$$\phi_1' = \frac{2C\phi_2' \pm \sqrt{4C^2\phi_2'^2 - 4AB\phi_2'^2}}{2A}$$

$$= \frac{2C\phi_2' \pm 2\sqrt{C^2 - AB}\phi_2'}{2A}$$

$$= \left[\frac{C}{A} \pm \frac{\sqrt{C^2 - AB}}{A} \right] \phi_2'$$

The warping compatibility factor γ can thus be obtained as

$$\gamma = \frac{C}{A} \pm \frac{\sqrt{C^2 - AB}}{A} \tag{5.32}$$

It is interesting to see that γ also signifies that the integral of the square of warping normal stress errors at the junction is zero. To show this, we have

$$\int \left[E\omega_1 \phi_1'' - E\omega_2 \phi_2'' \right]^2 ds = 0$$

which gives the quadratic equation

$$A\phi_1''^2 - 2C\phi_1'' \phi_2'' + B\phi_2''^2 = 0$$

Thus,

$$\phi_1'' = \gamma \, \phi_2''$$

The analysis of structures with junctions of dissimilar sections can now be carried out by using the continuity conditions (5.31).

As an example, the structure shown in Fig. 5.4 is considered. It consists of two box section portions joined to a central portion of channel section. The problem

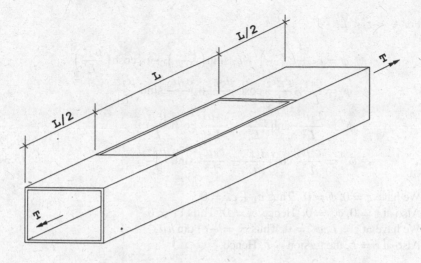

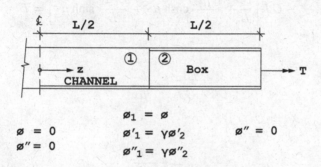

$\phi_1 = \phi$

$\phi = 0$ $\phi'_1 = \gamma\phi'_2$ $\phi'' = 0$

$\phi'' = 0$ $\phi''_1 = \gamma\phi''_2$

Fig. 5.4 Structure with junctions of dissimilar cross section

can be solved considering one-half of the structure with the boundary and junction conditions shown in Fig. 5.4. The solution of the problem is

For $0 < z < \frac{L}{2}$

$$\phi = c_1 + c_2\left(\frac{z}{L}\right) + c_3 \sinh\left(\frac{\mu_1 z}{L}\right) + c_4 \cosh\left(\frac{\mu_1 z}{L}\right)$$

$$\phi' = \frac{c_2}{L} + \frac{c_3\mu_1}{L}\cosh\frac{\mu_1 z}{L} + \frac{c_4\mu_1}{L}\sinh\frac{\mu_1 z}{L}$$

$$\phi'' = \frac{c_3\mu_1^2}{L^2}\sinh\frac{\mu_1 z}{L} + \frac{c_4\mu_1^2}{L^2}\cosh\frac{\mu_1 z}{L}$$

$$\phi''' = \frac{c_3\mu_1^3}{L^3}\cosh\frac{\mu_1 z}{L} + \frac{c_4\mu_1^3}{L^3}\sinh\frac{\mu_1 z}{L} \tag{5.33}$$

For $\frac{L}{2} < z < L$

$$\phi = c_5 + c_6\left(\frac{z}{L}\right) + c_7 \sinh\left(\frac{\mu_2 z}{L}\right) + c_8 \cosh\left(\frac{\mu_2 z}{L}\right)$$

$$\phi' = \frac{c_6}{L} + \frac{c_7\mu_2}{L}\cosh\frac{\mu_2 z}{L} + \frac{c_8\mu_2}{L}\sinh\frac{\mu_2 z}{L}$$

$$\phi'' = \frac{c_7\mu_2^2}{L^2}\sinh\frac{\mu_2 z}{L} + \frac{c_8\mu_2^2}{L^2}\cosh\frac{\mu_2 z}{L}$$

$$\phi''' = \frac{c_7\mu_2^3}{L^3}\cosh\frac{\mu_2 z}{L} + \frac{c_8\mu_2^3}{L^3}\sinh\frac{\mu_2 z}{L} \qquad (5.34)$$

We have $z = 0$, $\phi = 0$, Thus $c_1 + c_4 = 0$
Also at $z = 0$, $\phi'' = 0$, Hence $c_4 = 0$. Thus $c_1 = 0$
We have at $z = L$, $\phi'' = 0$. Thus $c_8 = -c_7 \tan h\mu_2$
Also at $z = L$, the torsion is T. Hence,

$$EI_{\Omega 2}\frac{\mu_2^3}{L^3}[c_7 \cosh \mu_2 + c_8 \sinh \mu_2]$$

$$- GJ_2\left[\frac{c_6}{L} + \frac{C_7\mu_2}{L}\cosh \mu_2 + \frac{C_8\mu_2}{L}\sinh \mu_2\right] = T$$

But

$$\frac{EI_{\Omega 2}}{GJ_2} = \frac{L^2}{\mu_2^2}$$

Hence,

$$c_6 = -\frac{TL}{GJ_2}$$

At $z = \frac{L}{2}$, the torsional moment obtained using Eqs. (5.33) and (5.34) must be the same, i.e.,

$$\left[EI_{\Omega 1}\phi''' - GJ_1\phi'\right]_{z=\frac{L}{2}} = \left[EI_{\Omega 2}\phi''' - GJ_2\phi'\right]_{z=\frac{L}{2}}$$

Hence,

$$EI_{\Omega 1}\frac{c_3\mu_1^3}{L^3}\cosh\frac{\mu_1}{2} - GJ_1\left[\frac{c_2}{L} + \frac{c_3\mu_1}{L}\cosh\frac{\mu_1}{2}\right]$$

$$= EI\frac{\mu_2^3}{L^3}\left[c_7\cosh\frac{\mu_2}{2} + c_8\sinh\frac{\mu_2}{2}\right]$$

$$- GJ_2\left[-\frac{T}{GJ_2} + \frac{c_7\mu_2}{L}\cosh\frac{\mu_2}{2} + \frac{c_8\mu_2}{L}\sinh\frac{\mu_2}{2}\right]$$

Using the notation

$$\frac{J_1}{J_2} = \delta$$

And

$$\alpha = \delta - \frac{1}{\gamma}$$

where γ is the warping compatibility factor. The foregoing equation becomes

$$\frac{c_3}{\gamma} \frac{\mu_1}{L} \cosh \frac{\mu_1}{2} - \frac{\alpha c_2}{L} = \frac{c_7 \mu_2}{L} \cosh \frac{\mu_2}{2} + \frac{c_8 \mu_2}{L} \sinh \frac{\mu_2}{2}$$

Compatibility relations (5.31) at the junction of the cross sections, i.e., at $z = \frac{L}{2}$, give

$$\frac{c_2}{2} + c_3 \sinh \frac{\mu_1}{2} = c_5 - \frac{TL}{2GJ_2} + c_7 \sinh \frac{\mu_2}{2} + c_8 \cosh \frac{\mu_2}{2}$$

$$\frac{c_2}{L} + \frac{c_3 \mu_1}{L} \cosh \frac{\mu_1}{2} = \gamma \left[-\frac{T}{GJ_2} + \frac{c_7 \mu_2}{L} \cosh \frac{\mu_2}{2} + \frac{c_8 \mu_2}{L} \sinh \frac{\mu_2}{2} \right]$$

$$\frac{c_3 \mu_1^2}{L^2} \sinh \frac{\mu_1}{2} = \gamma \left[\frac{c_7 \mu_2^2}{L^2} \sinh \frac{\mu_2}{2} + \frac{c_8 \mu_2^2}{L^2} \cosh \frac{\mu_2}{2} \right]$$

The foregoing equations can be solved to give

$$c_2 = -\frac{TL}{GJ_2 \left(\frac{1}{\gamma} + \alpha \right)}$$

$$c_7 =$$
$$\frac{TL\alpha \mu_1 \gamma \sinh \frac{\mu_1}{2}}{GJ_2 \, (1 + \alpha\gamma)\mu_2 \left[\mu_2 \tanh \mu_2 \cosh \frac{\mu_1}{2} \cosh \frac{\mu_2}{2} - \mu_2 \cosh \frac{\mu_1}{2} \sinh \frac{\mu_2}{2} - \mu_1 \sinh \frac{\mu_1}{2} \tanh \mu_2 \sinh \frac{\mu_2}{2} + \mu_1 \sinh \frac{\mu_1}{2} \cosh \frac{\mu_2}{2} \right]}$$

$$c_3 = \frac{c_7 \mu_2 \cosh \frac{\mu_2}{2} + c_8 \mu_2 \sinh \frac{\mu_2}{2} + \alpha c_2}{\frac{\mu_1}{\gamma} \cosh \frac{\mu_1}{2}}$$

$$c_5 = \frac{TL}{2GJ_2} + \frac{c_2}{2} + c_3 \sinh \frac{\mu_1}{2} - c_7 \sinh \frac{\mu_2}{2} - c_8 \cosh \frac{\mu_2}{2}$$

This completes the solution of the structure.

Fig. 5.5 Review problem 1

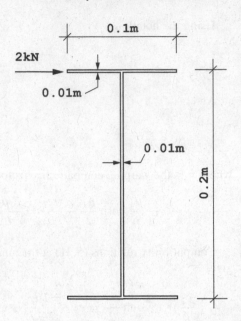

5.5.1   Other Warping Restraints

Apart from abrupt variations in cross section, other forms of warping restraints may also exist at internal points in the structure. Thin diaphragms do not restrain warping as their out-of-plane stiffness is small compared to their in-plane stiffness. They thus restrain only the distortion of the cross section. Diaphragms are thus outside the purview of internal warping restraining structures.

Transverse strips, small boxes in the transverse direction, etc., which tie tops of open sections at specific intervals constitute warping restraints at these points. The restraint arises from their lateral bending stiffness which contributes a restraint to the longitudinal (*warping*) displacement of the structure. The restraint can be taken into account in the analysis by defining a 'warping displacement spring' of stiffness k_Ω (units are Nm^3), so that the bimoment at the section introduced by the restraint is $k_\Omega \phi'$. The value of the warping displacement spring can be computed from the geometry and cross-sectional (*flexural*) properties of the transverse strip. The torsional analysis of the structure can now be carried out similar to beams on discrete elastic supports.

5.6   Review Problems

1. The section shown in Fig. 5.5 is the end section of a cantilever, one meter long. Determine the horizontal deflection of the loaded point as well as the maximum longitudinal stress at the fixed end. $E = 200$ GPa and $G = 70$ GPa.

2. An I beam of cross-sectional dimensions shown in Fig. 5.6 is fixed at one end and rigidly attached to a massive plate at the other end. For an allowable normal stress of 160 MPa, find the magnitude of the load P. $E = 72$ GPa and $G = 27.1$ GPa.

3. The structure shown in Fig. 5.1 is subjected to a torque $T = 0.2$ GNm. The span of the structure is 70 m. The cross section is multicellular with the following properties.

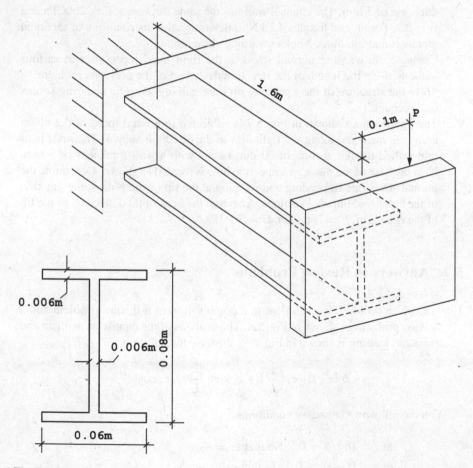

Fig. 5.6   Review problem 2

$$GJ = 1006 \text{ GNm}^2$$

$$EI_\Omega = 2450000 \text{ GNm}^4$$

Determine the distributions of torsional rotation, St. Venant and Vlasov contributions of torsions along the length in mixed torsion.

4.   The structure shown in Fig. 5.2 is subjected to a distributed torque of 0.0037143 GN in each half with opposing directions. Find the distributions of torsional rotation and St. Venant and Vlasov contributions of the torsion in mixed torsion. The cross section is the same as given in the previous example.

5.   In the foregoing example, if the warping of the structure is further restrained at the right end, determine the distributions of torsional rotations and torsions along the length.

6.   The structure shown in Fig. 5.4 has the channel portion 1.2 m long and each of the end boxes 0.6 m long. The size of the box is 0.4 m by 0.2 m, with a wall thickness of 3 mm. The channel wall has the same thickness. $E = 200$GPa and $\upsilon = 0.3$. For an end torque of 1 kNm, determine the distributions of torsional moments and rotations. Neglect warping shear strains.

7.   Compute the warping normal stress at the right hand top corner at various sections along the length of the structure described in the previous problem.

8.   Solve the structure of the foregoing problem talking warping shearing strains into account.

9.   The lighting mast shown in Fig. 5.7 is subjected to a wind force of 0.2 kN/m along the mast and along the right arm in the direction shown. The mast is of thin-walled tubular section of 60 mm radius with a wall thickness of 3 mm. Near the base of the mast, an inspection hole is located as shown. Determine the normal stress due to bending and warping at the tip of the hole at the junction of the holed and unholed sections. Also find the horizontal deflection of the tip of the rigid arm. $E = 206$ GPa, $G = 80$ GPa.

5.7   Answers to Review Problems

1.   The given problem is equivalent to a combination of a flexure problem and a torsion problem as shown in Fig. 5.8. The analysis of the cantilever with an end torsional loading is shown in Fig. 5.8. We have the solution

$$\phi = c_1 + c_2\left(\frac{z}{L}\right) + c_3 \sinh \frac{\mu z}{L} + c_4 \cosh \frac{\mu z}{L} \quad .$$

With the following boundary conditions

at $z = 0$, $\phi = 0$,   So that $c_1 = -c_4$

at $z = 0$, $\phi' = 0$,   So that $c_2 = -c_3\mu$

at $z = L$, $\phi'' = 0$,   So that $c_4 = -c_3 \tanh \mu$

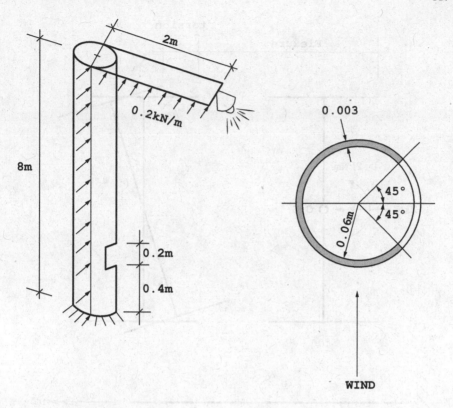

Fig. 5.7 Review problem 9

$$\text{at } z = L, \quad EI_\Omega\phi''' - GJ\phi' = T, \quad \text{So that } c_2 = -\frac{TL}{GJ}$$

The solution is then

$$\phi = \frac{TL}{\mu GJ}\left[\sinh\frac{\mu z}{L} - \frac{\mu z}{L} + \left\{1 - \cosh\frac{\mu z}{L}\right\}\tanh\mu\right]$$

The maximum rotation at $z = L$ is

$$\phi_m = \frac{TL}{\mu GJ}[\tanh\mu - \mu]$$

The flexural properties of the sections are

$$I = \frac{1}{6}(0.01)(0.1)^3 + \frac{1}{12}(0.2)(0.01)^3$$
$$= 0.1683(10^{-5}) \text{ m}^4$$

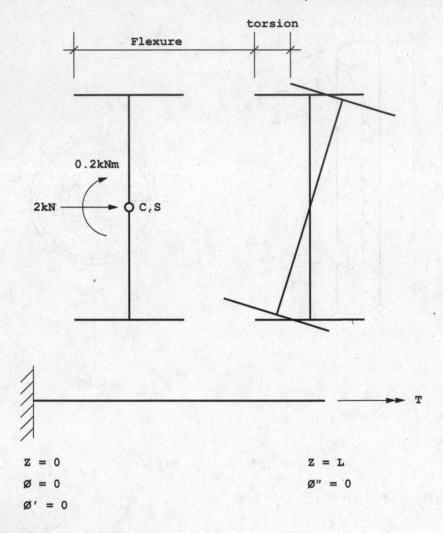

Fig. 5.8 Answer to review problem 1

The torsional properties of the section are

$$J = \frac{2}{3}(0.1)(0.01)^3 + \frac{1}{3}(0.2)(0.01)^3$$
$$= 0.1333(10^{-6}) \text{ m}^4$$

Unit warping at loaded point

$$\omega_n = \frac{bh}{4} = \frac{(0.1)(0.2)}{4} = 0.005 \text{ m}^2$$

Also

$$I_\Omega = \frac{th^5}{24}\xi^3 \quad \text{where } \xi = \frac{b}{h} = \frac{0.1}{0.2} = 0.5$$

$$= \frac{(0.01)(0.2)^5(0.5)^3}{24}$$

$$= 166.6667(10^{-10}) \text{ m}^6$$

Bending deflection of the loaded point

$$= \frac{2(10^{-6})(1)}{3(200)(0.1683)(10^{-5})} = 1.98(10^{-3}) \text{ m}$$

we have $d = \sqrt{\dfrac{EI_\Omega}{GJ}} = 0.597614$

$$\mu = \frac{L}{d} = \frac{1}{0.597614} = 1.67332$$

Thus,

$$\phi_m = \frac{2(10^{-7})(1)(0.7413308)}{(1.67332)(70)(0.1333)(10^{-6})} = 0.0094935 \text{ rad}$$

Horizontal deflection of loaded point due to torsion

$$= (0.1118)(0.0094935) = 1.06 \times 10^{-3} \text{ m}$$

Thus, the total horizontal deflection of the loaded point

$$= 1.98 + 1.06 = 3.04 \text{ mm}$$

Maximum bending stress at the fixed end

$$= \frac{2(10^{-3})(1)(0.05)}{(0.1683334)(10^{-5})} = 59.41 \text{ MPa}$$

Maximum normal stress due to warping

$$= E\omega_n\phi''$$

$$= E\omega_n \frac{c_4\mu^2}{L^2}$$

$$= \frac{TE\omega_n\mu \tanh \mu}{GJL}$$

$$= \frac{2(10^{-7})(200)(0.005)(1.67332)(0.9319892)}{(70)(0.1333)(10^{-6})(1)} \text{ GPa}$$

$$= 33.42(10^{-3}) \text{ GPa}$$

The maximum normal stress at the fixed end is thus

$$= 59.41 + 33.42 = 92.86 \text{ MPa}$$

2. The problem is equivalent to a combination of a flexure problem and a torsion
    problem. The torsion problem has the solution

$$\phi = c_1 + c_2\left(\frac{z}{L}\right) + c_3 \sinh\left(\frac{\mu z}{L}\right) + c_4 \cosh\left(\frac{\mu z}{L}\right)$$

With the following boundary conditions

$$\text{at } z = 0, \quad \phi = 0$$
$$\text{at } z = L, \quad \phi' = 0$$
$$\text{at } z = L, \quad EI_\Omega \phi''' - GJ\phi' = T$$

The constants can be determined as

$$c_2 = -\frac{TL}{GJ}$$

$$c_3 = \frac{TL}{\mu GJ}$$

$$c_4 = \frac{TL}{GJ\,\mu\,\sinh\mu}(1 - \cosh\mu)$$

$$c_1 = -c_4$$

For the given I selection, the following properties can be determined.

$$I = \frac{1}{12}(0.006)(0.068)^3 + \frac{1}{6}(0.06)(0.006)^3 + 2(0.06)(0.006)(0.037)^2$$

$$= 1.145056(10^{-6}) \text{ m}^4$$

$$J = \frac{2}{3}(0.06)(0.006)^3 + \frac{1}{3}(0.068)(0.006)^3$$

$$= 1.3536(10^{-8}) \text{ m}^4$$

$$\omega_n \text{ (tip)} = \frac{bh}{4} = \frac{(0.08)(0.06)}{4} = 0.0012 \text{ m}^2$$

$$I_\Omega = \frac{(0.006)(0.08)^5(0.75)^3}{24}$$

$$= 3.456(10^{-10}) \text{ m}^6$$

$$d = \sqrt{\frac{72(3.456)(10^{-10})}{27.1(1.3536)(10^{-8})}} = 0.26044936$$

$$\mu = \frac{L}{d} = \frac{1.6}{0.26044936} = 6.1432278$$

With load P expressed in kN, we have

$$c_4 = \frac{P(10^{-3})(0.1)(1.6)(1 - 232.77801)}{27.1(10^3)(1.3536)(10^{-8})(6.1432278)(232.77586)}$$

$$= -0.0706965\ P$$

The warping normal stress at the tip of the section at the fixed end

$$= E\omega_n \phi''$$

$$= (72)(10^3)(0.0012)\frac{0.0706965\ P(6.1432278)^2}{(1.6)^2}$$

$$= 90.046123P\ \text{MPa}$$

The flexural (*normal*) stress at the tip at the fixed end is

$$= \frac{1.6\ P(10^{-3})(0.04)}{1.145056(10^{-6})} = 55.892463P\ \text{MPa}$$

Hence, the total normal stress at the tip $= 145.93859P$ MPa
The value of the load P is thus

$$P = \frac{160}{145.93859} = 1.096\ \text{kN}$$

3. The solution of the problem is given by Eqs. (5.14) and (5.15). A computer program is written in FORTRAN IV for the solution and is presented in the following.

```
C234567890123456789012345678901234567890123456789 0
C          1         2         3         4         5
CALCULATIONS FOR MIXED TORSION -POINT TORQUE
      READ(25,10) AL,T,EI,GJ
  10  FORMAT(4F10.0)
      D=SQRT(EI/GJ)
      PMU=AL/D
      WRITE(26,20)D,PMU
  20  FORMAT(5X,5 (E20.6,5X))
```

```
   30  FORMAT(5X,")
       WRITE(26,30)
       C7= -T*AL*SINH (0.5*PMU)/(GJ*PMU*TANH (PMU))
       C3=C7*(1. - TANH (PMU))/(TANH (0.5*PMU))
       C2=-T*AL/(2.*GJ)
       C5=C2
       WRITE(26,20)C2,C3,C5,C7
       WRITE(26,30)
   ZL=0.0KIND=0
   100 CONTINUE
       IF(ZL.GT.0.51)GOTO200
       FI=C2*ZL+C3*SINH (PMU*ZL)
       FI1=C2/AL+C3*PMU*COSH (PMU*ZL)/AL
        FI2=C3*(PMU**2)*SINH (PMU*ZL)/(AL**2)
       FI3=C3*(PMU**3)*COSH*(PMU*ZL)/(AL**3)
       GOTO300
   200 FI=C5*(1.-ZL)+C7*(SINH (PMU*ZL)-TANH (PMU)*
       $COSH(PMU*ZL))
        FI1=-C5/AL+C7*(PMU*COSH(PMU*ZL)/AL-PMU*TANH
       $PMU)*SINH (PMU*ZL)/AL)
        FI2=C7*(SINH (PMU*ZL)*(PMU**2)/(AL**2)-(PMU**2)*
       $TANH(PMU)*COSH (PMU*ZL)/(AL**2))
        FI3=C7*(COSH(PMU*ZL)*(PMU**3)/(AL**3)-(PMU**3)
       $*TANH(PMU)*SINH(PMU*ZL)/(AL**3))
   300 WRITE(26,20)ZL,FI1,FI2,FI3
       WRITE(26,30)
       IF(KIND.EQ.1)GOTO500
       IF(ZL.LT.0.51.AND.ZL.GT0.49)KIND=1
        IF(KIND.EQ.1)GOTO 200
   500 ZL=ZL+0.1
       IF(ZL.GT.1.1)GOTO400 GOTO100
   400 WRITE(26,30)
       STOP
       END
```

The following results are obtained.

$\frac{z}{L}$	$-\phi \times 10^3$	$\phi' \times 10^4$	$\phi'' \times 10^6$	$\phi''' \times 10^7$	$EI_\Omega \phi'''$	$GJ\phi'$	T
0	0.0	−0.2065	0	0.3234	0.0792	−0.0208	0.10
0.1	0.1427	−0.1986	0.2271	0.3266	0.0800	−0.02	0.10
0.2	0.2743	−0.1746	0.4588	0.3365	0.0824	−0.0176	0.10
0.3	0.3833	−0.1341	0.6998	0.3531	0.0865	−0.0135	0.10
0.4	0.4580	−0.0763	0.9548	0.3768	0.0923	−0.0077	0.10
0.5^-		0		0.4081	0.1	0	0.10
0.5^+	0.4858	0	1.2291	−0.4081	−0.1	0	−0.10
0.6	0.4580	0.0763	0.9548	−0.3768	−0.0923	0.0077	−0.10
0.7	0.3833	0.1341	0.6998	−0.3531	−0.0865	0.0135	−0.10

(continued)

(continued)

$\frac{z}{L}$	$-\phi \times 10^3$	$\phi' \times 10^4$	$\phi'' \times 10^6$	$\phi''' \times 10^7$	$EI_\Omega \phi'''$	$GJ\phi'$	T
0.8	0.2743	0.1746	0.4588	−0.3365	−0.0824	0.0176	−0.10
0.9	0.1427	0.1986	0.2271	−0.3266	−0.08	0.02	−0.10
1.0	0	0.2065	0	−0.3234	−0.0792	0.0208	−0.10

In this example, the value of μ is 1.4184 which is small. Hence, the warping torsion predominates as seen in the foregoing table. The rotation of the central section obtained by pure St. Venant and by pure warping torsion analyses [Eqs. (5.5) and (5.8)] is compared in the following table with the actual *(combined torsion)* analysis value.

St. Venant	Vlasov	Mixed
3.4791×10^{-3}	0.5833×10^{-3}	0.4858×10^{-3}

4. The solution of the problem is given by Eqs. (5.24) through (5.29). A computer program is written for the solution and is presented in the following.

```
C2345678901234567890123456789012345678901234567890
C         1         2         3         4         5
CALCULATION FOR MIXED TORSION CARGO TORQUE
      READ(25,10)AL,T,EI,GJ
   10 FORMAT(4F10.5)
      D=SQRT(EI/GJ)
      PMU=AL/D
      WRITE(26,20)D,PMU
   20 FORMAT(5X,5(E20.6,5X))
   30 FORMAT(5X,'      ')
      WRITE(26,30)
      C6=-T*(AL**2)/GJ
      C7=2.0*T*(AL**2)*SINH(0.5*PMU)/(GJ*PMU*PMU)
      C8=1.0+2.0*SINH(PMU)*SINH(0.5*PMU)
      C8=-T*(AL**2)*C8/(GJ*COSH(PMU)*PMU*PMU)
      C1=-2.0*COSH(0.5*PMU)+1.0/COSH(PMU
     $+2.*TANH(PMU)* SINH (0.5*PMU)
      C1=T*(AL**2)*C1/(GJ*PMU*PMU)
      C4=-C1
      C5=C1+T*(AL**2)*(.25+2./(PMU*PMU)/GJ
      WRITE(26,20)C1,C5,C6,C7,C8
      WRITE(26,30)
      ZL=0.0
      KIND=0
  100 CONTINUE
      IF(ZL.GT.0.51)GOTO200
      FI=C1+C4*COSH(PMU*ZL)*0.5*T*ZL*AL*ZL*AL/GJ
      FI1=C4*PMU*SINH(PMU*ZL)/AL-T*ZL*AL/GJ
```

```
        FI2=C4*(PMU**2)*COSH(PMU*ZL)/(AL**2)-T/GJ
        FI3=C4*(PMU**3)*SINH(PMU*ZL)/(AL**3)
        GOTO 300
   200  FI=C5+C6*ZL+C7*SINH(PMU*ZL)+C8*COSH(PMU*ZL)+
        $0.5*T*(ZL** 2)*(AL**2)/GJ
        FI1=C6/AL+C7*PMU*COSH(PMU*ZL)/AL+C8*PMU
        $*SINH(PMU*ZL)/AL+T*ZL*AL/GJ
        FI2=C7*(PMU**2)*SINH(PMU*ZL)/(AL**2)+C8*(PMU**2)
        $*COSH(PMU*ZL)/(AL**2)+T/GJ
        FI3=C7*(PMU**3)*COSH(PMU*ZL)/(AL**3)+C8*(PMU**3
        $*SINH(PMU*ZL)/(AL**3)
   300  WRITE(26,20)ZL,FI,FI1,FI2,FI3
        WRITE(26,30)
        IF(KIND.EQ.1)GOTO500
        IF(ZL.LT.0.51.AND.ZL.GT.0.49)KIND=1
        IF(KIND.EQ.1)GOTO200
   500  ZL=ZL+0.1
        IF(ZL.GT.1.1)GOTO400
        GOTO100
   400  WRITE(26,30)
        STOP
        END
```

Following results are obtained.

$\frac{z}{L}$	$-\phi \times 10^3$	$-\phi' \times 10^4$	$\phi'' \times 10^6$	$\phi''' \times 10^7$	$EI_\Omega\phi'''$	$-GJ\phi'$	T
0	0	0	−1.1179	0	0	0	0
0.1	0.0273	0.0776	−1.0919	0.0742	0.0182	0.0078	0.0260
0.2	0.1079	0.1516	−1.0136	0.1500	0.0368	0.0153	0.0521
0.3	0.2379	0.2183	−0.8812	0.2287	0.0560	0.0220	0.0780
0.4	0.4108	0.2737	−0.6923	0.3121	0.0765	0.0275	0.1040
0.5	0.6175	0.3138	−0.4428	0.4018	0.0984	0.0316	0.1300
0.6	0.8458	0.3359	−0.2023	0.2866	0.0702	0.0338	0.1040
0.7	1.0845	0.3439	−0.0403	0.1771	0.0434	0.0346	0.0780
0.8	1.3253	0.3433	0.0465	0.0713	0.0175	0.0345	0.0521
0.9	1.5642	0.3391	0.0599	−0.0331	−0.0081	0.0341	0.0260
1.0	1.8005	0.3366	0	−0.1382	−0.0339	0.0339	0

Because of the small value of μ, the warping torsion predominates. A comparison of the end rotation obtained by the mixed torsion analysis is made in the following with those obtained by pure St. Venant and pure Vlasov analyses.

St. Venant	Vlasov	Mixed
4.53×10^{-3}	3.21×10^{-3}	1.8005×10^{-3}

5.   The solution of the problem is defined by Eq. (5.30a). A computer program is
     written for the solution and is presented in the following.

```
C234567890123456789012345678901234567890123456789 0
C         1         2         3         4         5
CALCULATIONS FOR MIXED TORSION CARGO TORQUE /ENDS NO WARPING
      READ(25,10)AL,T,E,GJ
  10  FORMAT(4F10.0)
      D=SQRT(EI/GJ)
      PMU=AL/D
      WRITE(26,20)D,PMU
  20  FORMAT(5X,5(E20.6,5X))
  30  FORMAT(5X,'      ')
      WRITE(26,30)
      C6=T*(AL**2)/GJ
      C7=2.*T*(AL**2)*SINH(.5*PMU)/(PMU*PMU*GJ)
      C8=SINH(0.5*PMU)/TANH(PMU)
      C8=2.*T*(AL**2)*C8/(PMU*PMU*GJ)
      C1=SINH(0.5*PMU)/TANH (PMU)*SINH(0.5*PMU)
     $*TANH(0.5*PMU)-1./COSH(0.5*PMU)
      C1=2.*T*(AL**2)*C1/(PMU*PMU*GJ)
      C4=-C1
      C5=C1+T*(AL**2)*(1.+8./(PMU*PMU))/(4.*GJ)
      WRITE(26,20)C1,C5,C6,C7,C8
      WRITE(26,30)
      ZL=0.0
      KIND=0
 100  CONTINUE
      IF(ZL.GT.0.51)GOTO200
      FI=C1+C4*COSH(PMU*ZL)*0.5*T*ZL*AL*ZL*AL/GJ
      FI1=C4*PMU*SINH(PMU*ZL)/AL*T*ZL*AL/GJ
      FI2=C4*(PMU**2)*COSH(PMU*ZL)/(AL**2)-T/GJ
      FI3=C4*(PMU**3)*SINH(PMU*ZL)/(AL**3)
      GOTO300
 200  FI=C5+C6*ZL+C7*SINH(PMU*ZL)
     $+C8*COSH(PMU*ZL)+0.5*T*(ZL**2)*(AL**2)/GJ
 FI1=C6/AL+C7*PMU*COSH(PMU*ZL)/AL+C8*PMU*SINH
     $(PMU*ZL)/AL+T*ZL*AL/GJ
      FI2=C7*(PMU**2)*SINH(PMU*ZL)/(AL**2)
     $+C8*(PMU**2)*COSH(PMU*ZL)/(AL**2)+T/GJ
      FI3=C7*(PMU**3)*COSH(PMU*ZL)/(AL**3)
     $+C8*(PMU**3)*SINH(PMU*ZL)/(AL**3)
 300  WRITE(26,20)ZL,FI,FI1,FI2,FI3
      WRITE(26,30)
      IF(KIND.EQ.1)GOTO500
      IF(ZL.LT.*.51.AND.ZL.GT.*.49)KIND=1
      IF(KIND.EQ.1)GOTO200
 500  ZL=ZL+0.1
```

```
    IF(ZL.GT.1.1)GOTO400
    GOTO100
400 WRITE(26,30)
    STOP
    END
```

The following results are obtained.

$\frac{z}{L}$	$-\phi \times 10^3$	$-\phi' \times 10^4$	$\phi'' \times 10^6$	$\phi''' \times 10^7$	$EI_\Omega\phi'''$	$-GJ\phi'$	T
0	0	0	−0.7670	0	0	0	0
0.1	0.0187	0.0530	−0.7376	0.0844	0.0207	0.0053	0.0263
0.2	0.0732	0.1019	−0.6485	0.1704	0.0417	0.0103	0.0520
0.3	0.1593	0.1424	−0.4982	0.2599	0.0637	0.0143	0.0780
0.4	0.2696	0.1701	−0.2834	0.3546	0.0869	0.0171	0.1040
0.5	0.3934	0.1804	0	0.4565	0.1118	0.0181	0.1299
0.6	0.5173	0.1701	0.2834	0.3546	0.0869	0.0171	0.1040
0.7	0.6275	0.1424	0.4982	0.2599	0.0637	0.0143	0.0780
0.8	0.7136	0.1019	0.6485	0.1704	0.0417	0.0103	0.0520
0.9	0.7682	0.0530	0.7376	0.0844	0.0207	0.0053	0.0263
1.0	0.7869	0	0.7670	0	0	0	0

A comparison of the end rotation in pure St.Venant, pure Vlasov, and mixed torsion analyses is shown in the following

St. Venant	Vlasov	Mixed
4.53×10^{-3}	0.94×10^{-3}	0.79×10^{-3}

6. For the channel and box sections, the sectional properties can be calculated based on the details presented in Chap. 3. The normalized warping functions are shown in Fig. 5.9. We have for channel section:

$$I_\Omega = 2.8 \times 10^{-7} \text{ m}^6$$
$$J = 7.2 \times 10^{-9} \text{ m}^4$$
$$EI_\Omega = 58800 \text{ Nm}^4$$
$$GJ = 581.5385 \text{ Nm}^2$$

For box section:

$$I_\Omega = 5.3333 \times 10^{-8} \text{ m}^6$$

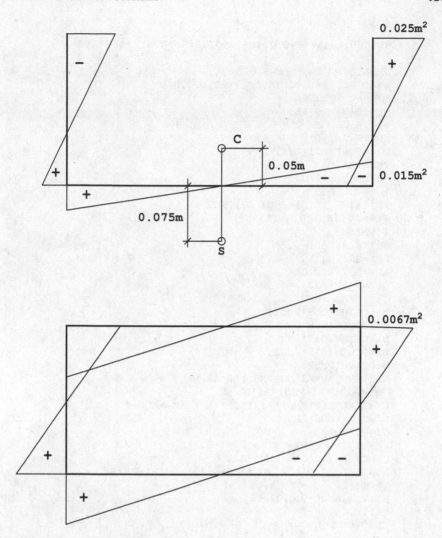

Fig. 5.9 Normalized warping functions of problem 6

$$J = 0.000064 \ \text{m}^4$$
$$EI_\Omega = 11200 \ \text{Nm}^4$$
$$GJ = 5169231 \ \text{Nm}^2$$

The warping compatibility factor for the junction can be obtained as $\frac{1}{3}$ using Eq. (5.32).

A computer program is written for the solution given by Eqs. (5.33) and (5.34). The program is presented in the following.

```
C2345678901234567890123456789012345678901234567890
C         1         2         3         4         5
      IMPLICIT REAL*8(A-H,0-Z)
      READ(25,1)AL,T,EI1,GJ1,EI2,GJ2,GAM
    1 FORMAT(7F10.0)
      WRITE(26,20)AL,T
      WRITE(26, 20)EI1,GJ1,EI2,GJ2,GAM
      WRITE(26,30)
      D1=DSQRT(EI1/GJ1)
      PMU1=AL/D1
      D2=DSRT(EI2/GJ2)
      PMU2=AL/D2
      WRITE(26,20)D1,PMU1,D2,PMU2
   20 FORMAT(5X,5(D20.I2,5X))
   30 FORMAT(5X,'              ')
      WRITE(26,30)
      DEL=GJ1/GJ2
      ALP=DEL-1.0/GAM
      C6=-T*AL/GJ2
      C2=-T*AL/(GJ2*(1.0/GAM+ALP))
      A1=DSINH(0.5*PMU1)
      A2=DCOSH(0.5*PMU1)
      B1=DSINH(0.5*PMU2)
      B2=DCOSH(0.5*PMU2)
      U=DTANH(PMU2)
      RHS=PMU2*U*A2*B2-PMU2*A2*B1-
     $PMU*A1*U*B1+PMU1*A1*B2
      C7=T*AL*ALP*PMU1*A1/(GJ2*(1./GAM +ALP))
      C7=C7/(PMU2*RHS)
      C8=-C7*U
      RHS=C7*PMU2*B2+C8*PMU2*B1+C2*ALP
      C3=RHS/(1.0/GAM*PMU1*A2)
      C5=T*AL/(2.0*GJ2)+0.5*C2+C3*A1-C7*B1-C8*B2
      WRITE(26,20) C2,C3,C5
      WRITE(26,20) C6,C7,C8
      WRITE(26,30)
      ZL=0.0
      KIND=0
  100 CONTINUE
      IF(ZL.GT.0.51)GOTO200
      FI1=C2*ZL+C3*DSINH(PMU1*ZL)
      FI1=C2/AL+C3*PMU1*DCOSH(PMU1*ZL)/AL
      FI2=C3*(PMU1**2)*DSINH(PMU1*ZL)/(AL**2)
      FI3=C3*(PMU1**3)*DCOSH(PMU1*ZL)/(AL**3)
      GOTO300
  200 FI=C5+C6*ZL+C7*DSINH(PMU2*ZL)
     $+C8*DCOSH(PMU2*ZL)
      FI1=C6/AL+C7*PMU2*DCOSH(PMU2*ZL)/AL+C8*PMU2*
     $DSINH(PMU2*ZL)/ALFI2=C7*PMU2**2*DSINH
     $(PMU2*ZL)/AL**2+C8*(PMU2**2)*DCOSH(PMU2*ZL)
     $/AL**2
```

```
        FI3=C7*(PMU2**3)*DCOSH(PMU2*ZL)/(AL**3)
        $+C8*(PMU2**3)*DSINH(PMU2*ZL)/AL**3
300 WRITE(26,20)ZL,FI,FI1,FI2,FI3
        WRITE(26,30)
        IF(KIND.EQ.1)GOTO500
        IF(ZL.LT.0.51ANDZL.GT.0.49)KIND=1
        IF(KIND.EQ.1)GOTO200
500 ZL=ZL+0.1
        IF(ZL.GT.1.1)GOTO400
        GOTO100
400 WRITE(26,30)
        STOP
        END
```

Following results are obtained.

$\frac{z}{L}$	$-\phi \times 10^2$	$-\phi' \times 10^2$	$\phi'' \times 10^2$	$\phi''' \times 10$	$EI_\Omega \phi'''$	$-GJ\phi'$	T
0	0	0.3594	0	0.1697	998	2	1000
0.1	0.0426	0.3472	0.2037	0.1697	998	2	1000
0.2	0.0824	0.3106	0.4073	0.1698	998	2	1000
0.3	0.1162	0.2495	0.6111	0.1698	998	1.5	999.5
0.4	0.1412	0.1639	0.8149	0.1699	999	1	1000
0.5^-	0.1546	0.0539	1.0189	0.1700	999.6	0.3	1000
0.5^+	0.1546	0.1616	3.0566	−6.5667	−7355	8353	1000
0.6	0.1630	0.0301	0.2319	−0.4973	−557	1556	1000
0.7	0.1658	0.0201	0.0183	−0.0391	−44	1039	995
0.8	0.1681	0.0183	0	0	0	1000	1000
0.9	0.1694	0.0183	0	0	0	1000	1000
1.0	0.1709	0.0183	0	0	0	1000	1000

It is seen that in the channel portion, the warping torsion predominates, whereas in the box portion, the St. Venant torsion is dominant.

7. The warping normal stress can be conveniently computed as shown in the following table.

$\frac{z}{L}$	ϕ''	ω_n	$\sigma_w = E\omega_n \phi''$ (MPa)
0	0	0.025	0
0.1	0.002037	0.025	10.2
0.2	0.004073	0.025	20.4
0.3	0.006111	0.025	30.6
0.4	0.008149	0.025	40.7

(continued)

(continued)

$\frac{z}{L}$	ϕ''	ω_n	$\sigma_w = E\omega_n\phi''$ (MPa)
0.5	0.010189	0.025	50.9
0.5	0.030566	0.0067	41.0
0.6	0.002319	0.0067	3.1
0.7	0.000183	0.0067	0.2

8. The warping shear strains are taken into account in Benscoter's theory. The differential equation can be written as

$$E\left(\frac{I_\Omega}{f_s}\right)\phi''' - GJ\phi' = T$$

in which f_s is the warping shear parameter given by

$$f_s = 1 - \frac{J}{I_h}$$

For open sections $f_s = 1.0$

For the box section of the example, f_s can be computed as $\frac{1}{9}$. The torsional analysis can be carried out using a value of 9(11200) for the warping torsional rigidity of the box section portion in the computer program. Following are the results.

$\frac{z}{L}$	$-\phi \times 10^2$	$-\phi' \times 10^2$	$\phi'' \times 10^2$	$\phi''' \times 10$	$EI_\Omega\phi'''$	$-f_sGJ\phi'$	f_s T
0	0	0.4541	0	0.1696	997	3	1000
0.1	0.0540	0.4420	0.2035	0.1696	997	3	1000
0.2	0.1051	0.4053	0.4071	0.1697	998	2	1000
0.3	0.1503	0.3442	0.6108	0.1697	998	2	1000
0.4	0.1867	0.2587	0.8145	0.1698	998	2	1000
0.5⁻	0.2114	0.1487	1.0183	0.1699	999		1000
0.5⁺	0.2114	0.4461	3.0549	-2.1885	-2451	2562	111
0.6	0.2481	0.2002	1.2925	-0.9275	-1040	1150	110
0.7	0.2650	0.0963	0.5447	-0.3946	-442	553	111
0.8	0.2735	0.0528	0.2245	-0.1715	-192	303	111
0.9	0.2787	0.0355	0.0806	-0.0830	-93	204	111
1.0	0.2826	0.0310	0	-0.0596	-67	178	111

The warping normal stresses at right hand top corner are

$\frac{z}{L}$	ϕ''	ω_n	σ_w (MPa)
0	0	0.025	0
0.1	0.002035	0.025	10.2
0.2	0.004071	0.025	20.4
0.3	0.006108	0.025	30.5
0.4	0.008145	0.025	40.7
0.5	0.010183	0.025	50.9
0.5	0.030549	0.0067	40.9
0.6	0.012925	0.0067	17.3
0.7	0.005447	0.0067	7.3
0.8	0.002245	0.0067	3.0
0.9	0.000806	0.0067	1.1
1.0	0	0.0067	0

9. The warping of a hollow circular section is negligible. Hence, the warping is confined to only the 0.2 m long open circular section. The section can be considered fixed against warping at its ends as shown in Fig. 5.10. The solution of the warping problem is

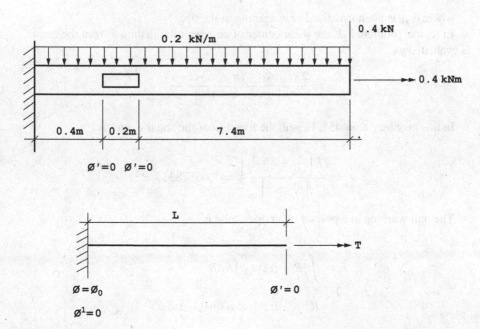

Fig. 5.10 Review problem 9

$$\phi = c_1 + c_2\left(\frac{z}{L}\right) + c_3 \sinh\frac{\mu z}{L} + c_4 \cosh\frac{\mu z}{L}$$

With the conditions
At $z = 0$, $\phi = \phi_0$ (the St. Venant rotation of 0.4 m ortion)
Hence, $c_1 + c_4 = \phi_0$
at $z = 0$, $\phi' = 0$. Thus, $c_2 = -c_3\mu$
at $z = L$, $\phi' = 0$. Hence,

$$c_4 = \frac{c_2(\cosh\mu - 1)}{\mu \sinh\mu}$$

Any where
$EI_\Omega\phi''' - GJ\phi' = T$ which gives

$$c_2 = -\frac{TL}{GJ}$$

The warping normal stress at the tip of the open section 0.6m from ground is thus

$$\sigma_w = E\omega_{n1}\frac{c_4\mu^2}{L^2}$$

where ω_{n1} is the normalized unit warping at the tip.
In review problem 2.3, the shear center of an open tube distant a from the center
is evaluated as

$$a = \frac{2R[\sin\alpha + (\pi - \alpha)\cos\alpha]}{\left[\pi - \alpha + \frac{\sin 2\alpha}{2}\right]}$$

In this problem, $\alpha = 45°$. Hence, the location of the shear center is

$$a = \frac{2R\left[\frac{1}{\sqrt{2}} + \frac{3\pi}{4}\frac{1}{\sqrt{2}}\right]}{\left[\frac{3\pi}{4} + \frac{1}{2}\right]} = 1.66178311\ R$$

The unit warping at a point θ in an open tube is

$$\omega = \int_\alpha^\theta [R + a\cos\beta]Rd\beta$$

$$= R^2(\theta - \alpha) + Ra(\sin\theta - \sin\alpha)$$

Now

$$\int \omega dA_s = \int\limits_{\alpha}^{2\pi-\alpha} \left[R^2(\theta - \alpha) + Ra(\sin\theta - \sin\alpha)\right](Rd\theta)(t)$$

$$= 2R^2t(\pi - \alpha)[R(\pi - \alpha) - a\sin\alpha]$$

The normalized unit warping at any point θ is given by

$$\omega_n = \omega - \frac{\int \omega dA_s}{A_s}$$

$$= R^2(\theta - \alpha) + Ra(\sin\theta - \sin\alpha)$$

$$- \frac{2R^2t(\pi - \alpha)[R(\pi - \alpha) - a\sin\alpha]}{2Rt(\pi - \alpha)}$$

$$= R^2(\theta - \pi) + Ra\sin\theta$$

The normalized unit warping at the tip $(\theta = \alpha)$ in this example is

$$\omega_{n1} = -\frac{3\pi}{4}R^2 + \frac{Ra}{\sqrt{2}}.$$

$$= -1.18113638\,R^2$$

$$= -1.18113638(0.06)^2 = -0.00425209\text{ m}^2$$

The warping moment of inertia can be evaluated as

$$I_\Omega = \int \omega_n^2\, dA_s$$

$$= \int\limits_{\alpha}^{2\pi-\alpha} \left[R^2(\theta - \pi) + Ra\sin\theta\right]^2 Rt\,d\theta$$

$$= Rt \int\limits_{\alpha}^{\pi-\alpha} \left[R^4(\theta - \pi)^2 + 2R^3a[\theta\sin\theta - \pi\sin\theta] + R^2a^2\sin^2\theta\right]d\theta$$

$$= Rt \left[\frac{R^4(\theta - \pi)^3}{3} + 2R^3a\{\sin\theta - \theta\cos\theta + \pi\cos\theta\} \right.$$

$$\left. + \frac{R^2a^2}{2}\left\{\theta - \frac{\sin 2\theta}{2}\right\} \right]_{\alpha}^{2\pi-\alpha}$$

$$= \frac{R^5t}{3}\left[(\pi - \alpha)^3 - (\alpha - \pi)^3\right] + 2R^4ta\{-2\sin\alpha - 2(\pi - \alpha)\cos\alpha\}$$

$$+ \frac{R^3a^2t}{2}\{2(\pi - \alpha) + \sin 2\alpha\}$$

$$= \frac{2R^5t}{3}(\pi - \alpha)^3 - 4R^4ta\{\sin\alpha + (\pi - \alpha)\cos\alpha\}$$

$$+ R^3a^2t\{\pi - \alpha + \sin\alpha\cos\alpha\}$$

$$= R^5 t \left[\frac{2}{3} R^2 (\pi - \alpha)^3 + a(\pi - \alpha)[a - 4R\cos\alpha] + a\sin\alpha[a\cos\alpha - 4R] \right]$$

For the current problem $\alpha = 45°$ and $a = 1.6617831\ R$.
Hence,

$$I_\Omega = (0.06)^5 (0.003)[8.72051532 - 4.56797622 - 3.3194708]$$
$$= 1.9433815(10^{-9})\ m^6$$

Also the torsion constant of the open section is

$$J = \frac{1}{3}[2R(\pi - \alpha)]t^3$$

With $\alpha = \frac{\pi}{4}$, we get

$$J = \frac{\pi (0.06)(0.003)^3}{2} = 2.54469 \times 10^{-9}\ m^4$$

Thus,

$$d = \sqrt{\frac{206(1.9433815)10^{-9}}{80(2.54469)10^{-9}}} = 1.4023299$$

$$\mu = \frac{0.2}{1.4023299} = 0.1426197$$

Hence,

$$c_2 = -\frac{0.4(10^{-3})(0.2)}{80(10^3)(2.54469)10^{-9}} = -0.3929751$$

$$c_4 = \frac{(-0.3929751)(0.0101874)}{(0.1426197)(0.1431038)} = -0.1961542$$

The warping normal stress at the tip of the hole at the junction is thus

$$\sigma_W = 206(10^3)(0.00425209)\frac{(0.1961542)(0.1426197)^2}{(0.2)^2}$$

$$= 87\ MPa$$

The moment of inertia for bending

$$I = R^3 t \left[\pi - \alpha + \frac{\sin 2\alpha}{2} \right]$$

For this example,

$$I = (0.06)^3(0.003)\left[\frac{3\pi}{4} + \frac{1}{2}\right] = 1.850814(10^{-6}) \text{ m}^4$$

The bending normal stress at the tip of the open section is

$$\sigma_b = \frac{[0.4(7.4) + (0.2)(7.4)(3.7)]\left[0.06/\sqrt{2}\right]}{1.850814(10^{-6})} \text{ kPa}$$
$$= 193(10^3) \text{ kPa} = 193 \text{ MPa}$$

The torsional rotation of 0.4 m portion is

$$\phi_0 = -\frac{(0.4)(10^{-3})(0.4)}{80(10^3)\left[2\pi(0.06)^3(0.003)\right]}$$
$$= -0.491219(10^{-3}) \text{ rad}$$

The torsional rotation at the top of the mast is

$$\phi_m = \phi_0 + c_4(\cosh\mu - 1) + c_3(\sinh\mu - 1) - \frac{(0.4)(10^{-3})(7.4)}{80(10^3)(4.0715)10^{-6}}$$
$$= [-0.491219 - 0.6644095 - 9.0876]10^{-3}$$
$$= -10.2434(10^{-3})$$

The deflection of the tip of the rigid arm due to the torsion of the tube mast is

$$= (2.0997069)(10.2432)10^{-3}$$
$$= 0.02151 \text{ m}$$

The deflection of the tip of the rigid arm due to the bending of the tubular mast

$$= \frac{(0.2)(10^{-3})(8)^4}{8(206)(10^3)(2.035752 \times 10^{-6})} + \frac{(0.4)10^{-3}(8)^3}{3(206 \times 10^3)(2.035752)(10^{-6})}$$
$$= 0.4069645 \text{ m}$$

The total deflection of the tip of the rigid arm is thus

$$= 407 + 22 = 429 \text{ mm}$$

Reference

1. Rajagopalan, K.: Warping torsional analysis of containerships. In: Proceedings of the Second Indian Conference in Ocean Engineering, CWPRS, Pune (1983)

Chapter 6
Finite Element Analysis of Thin-Walled Structures

6.1 Introduction

The use of finite elements for the torsional analysis of thin-walled structure is shown in this chapter. In the finite element analysis, the structure is modeled as an assembly of finite elements. The force–deflection characteristics of the structure are obtained by combining the force–deflection characteristics of the finite elements meeting at a joint to satisfy joint equilibrium. To analyze the St. Venant torsion problem, the structure is modeled with St. Venant torsion elements. Similarly, pure warping torsional analysis can be carried out by assembling the structure with Vlasov torsion elements. Mixed torsion problems can be tackled using mixed torsion finite elements. The use of finite elements is also illustrated for the determination of the warping properties such as normalized unit warping at the junctions of plates in a multicellular thin-walled section.

6.2 Stiffness Matrix

Any piece of a structure such as that shown in Fig. 6.1 (however small it may be) has infinite material points in it, and at all these points deflections, strains, and stresses are to be determined. In the displacements method, the piece is assumed to have only a finite degrees of freedom (six in the example shown in Fig. 6.1) and a function defining the shape of the entire structural piece is assumed so that the deflection at any point in the piece can be written in terms of these assumed degrees of freedom. Such functions are called shape functions. The degrees of freedom are called generalized coordinates. With this assumption, the force–deflection property of the structural piece is determined by relating the forces in the generalized coordinates to the displacements in the generalized coordinates. The relationship can of course be derived theoretically using the constitutive equation of the material.

© The Author(s), under exclusive license to Springer Nature Singapore Pte Ltd. 2022   137
K. Rajagopalan, *Torsion of Thin Walled Structures*,
https://doi.org/10.1007/978-981-16-7458-7_6

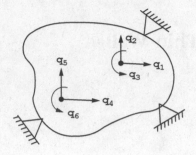 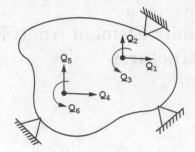

Fig. 6.1 Generalized forces and displacement

Any generalized force Q_i in Fig. 6.1, for example, can be written in terms of the generalized displacements q_1 to q_6 as follows

$$Q_1 = k_{11}q_1 + k_{12}q_2 + \cdots + k_{16}q_6$$

$$\vdots$$

$$Q_6 = k_{61}q_1 + k_{62}q_2 + \cdots + k_{66}q_6$$

which can be written as

$$\begin{Bmatrix} Q_1 \\ Q_2 \\ Q_3 \\ Q_4 \\ Q_5 \\ Q_6 \end{Bmatrix} = \begin{bmatrix} k_{11} & k_{12} & k_{13} & k_{14} & k_{15} & k_{16} \\ k_{21} & k_{22} & k_{23} & k_{24} & k_{25} & k_{26} \\ k_{31} & k_{32} & k_{33} & k_{34} & k_{35} & k_{36} \\ k_{41} & k_{42} & k_{43} & k_{44} & k_{45} & k_{46} \\ k_{51} & k_{52} & k_{53} & k_{54} & k_{55} & k_{56} \\ k_{61} & k_{62} & k_{63} & k_{64} & k_{64} & k_{66} \end{bmatrix} \begin{Bmatrix} q_1 \\ q_2 \\ q_3 \\ q_4 \\ q_5 \\ q_6 \end{Bmatrix} \tag{6.1}$$

Or simply

$$\{Q\} = [k]\{q\}$$

The stiffness matrix $[k]$ is symmetrical for linear structures because of the reciprocal theorem. Thus $k_{ij} = k_{ji}$. Any stiffness coefficient k_{ij} has units of force per unit length if it multiplies a deflection to give a force. It will have units of force if it multiplies a deflection to give a force. Similarly, it will have the unit of force if it multiplies a rotation to give moment.

The term generalized forces will mean forces or moments. Likewise the term generalized displacements may denote either deflections or rotations. The stiffness matrix of a structural piece is nothing but the relationship between the generalized forces and the generalized displacements, associated with the assumed set of generalized coordinates. If the shape function used to derive the stiffness matrix coincides with the actual displacements profile of the structure under the generalized forces, the stiffness matrix will be exact. Otherwise, it will be approximate. With exact stiffness matrices, larger finite elements (pieces) can be used in the structural analysis. If the stiffness matrix is approximate, a large number of smaller finite elements are needed to analyze the structure satisfactorily. The stiffness matrices of one-dimensional members can be derived in an exact manner. This is the reason why each member of a plane truss or a space truss or a plane frame or a space frame is taken as a single finite element. Subdivision of each member into smaller finite elements is unnecessary. Loads on the member (not occurring at the generalized coordinates) are taken into account by a two-stage principle to be enunciated in one of the following articles. Compared to the analysis of skeletal structures composed of one-dimensional members, the analysis of plates, shells, and solids are to be done using a large number of plate, shell, or solid elements as the stiffness matrices of these finite elements are difficult to derive in an exact manner. The two-stage principle is of course a general one and applies to these structures as well, but it is not used since a large number of finite elements are already employed.

The meaning of any column (or any row since the matrix is symmetrical) of the stiffness matrix can be explained as follows. With reference to Fig. 6.1 assume that six persons P1, P2, P3, P4, P5, P6 are given jacks which can apply forces or moments. The persons P3 and P6 can apply moments only; other can apply forces only at their own generalized coordinates. Suppose that these people are asked to get a deflected shape of the structural piece involving a unit displacement in generalized coordinate '1' and no displacements in any of the remaining generalized coordinates. To achieve this condition, P1 will start applying a force to induce a unit displacement in his own generalized coordinate '1' and all others (P2 to P6) will also start applying generalized forces to nullify the generalized displacements at their own coordinates. If we collect the forces offered by these people in order, it will form the first column of the stiffness matrix. This simple mental exercise though offered to illustrate the meaning of the stiffness matrix can actually be performed in a laboratory to derive the stiffness matrix of a finite element experimentally.

6.3 Structural Analysis

The force–deflection properties of a finite element are given by the element stiffness matrix. The structure is composed of a number of finite elements connected at the nodes of the structure. The degrees of freedom at the nodes of the structure are the structure degrees of freedom. These coincide with the element generalized coordinates if the stiffness matrix of the element refers to forces and displacements at a set of element generalized coordinates in the direction of the structure freedoms (If the stiffness matrix of the finite element in any other set of generalized coordinates is available, it can be transformed to give the stiffness which will refer to the structure coordinate directions). The structure stiffness matrix can then be built by merely assembling the element stiffness matrices. It will relate the forces and displacements at the generalized coordinates of the structure. The assembly means the satisfaction

of the equations of equilibrium at the joints (nodes) of the structure. The generalized forces of the structure have the meaning of applied loads. For these loads and
for boundary conditions on some of the structure freedoms, the remaining structure
freedoms are solved. Since the structure freedoms are also the element freedoms,
the generalized forces of the element can be determined by multiplying the element
freedoms by the element stiffness matrix. The stresses at all points of the elements
can be calculated from the force distribution. This is the basis of the finite element
method and is the procedure which applies to any type of structure modeled with any
type of finite element. The only needed entity is the stiffness matrix of pieces of the
structure (finite elements) in a form that refers to the structure directions so that the
elements can be directly connected and the structure solved for the given loads and
boundary conditions [1].

6.4   Expression for Stiffness Matrix

A finite element with generalized displacements $\{q\}$ is considered. These are the
degrees of freedom of the finite element and the displacements $\{u\}$ at any other point
can be written in terms of these generalized displacements using interpolation or
displacement shape. Thus,

$$\{u\} = [N]\{q\} \tag{6.2}$$

The strains $\{\epsilon\}$ at any point in the finite element can be obtained by appropriate
differentiation of the displacements $\{u\}$ as given by the appropriate strain–displacement relations. Hence, we can write

$$\{\epsilon\} = [B]\{q\} \tag{6.3}$$

where $[B]$ is obtained by differentiating the shape function matrix $[N]$. The matrix
$[B]$ is called the strain–displacement matrix.

The stresses $\{\sigma\}$ at any point are related to the strains by the constitutive relations
of the material. Thus

$$\{\sigma\} = [D]\{\epsilon\} \tag{6.4}$$

where $[D]$ is the material stiffness matrix.

The potential energy of the element can be written using the principle of virtual
displacement as

$$\pi = \int \delta\{\epsilon\}^{\mathrm{T}}\{\sigma\}\mathrm{d}V - \delta\{q\}^{\mathrm{T}}\{Q\}$$

$$= \int \delta\{q\}^{\mathrm{T}}[B]^{\mathrm{T}}[D][B]\{q\}\mathrm{d}V - \delta\{q\}^{\mathrm{T}}\{Q\}$$

The potential energy is a minimum for element equilibrium. Hence,

$$\frac{\partial \pi}{\partial \{q\}^{\mathrm{T}}} = 0$$

$$\left[\int [B]^{\mathrm{T}} [D] [B] \mathrm{d}V \right] \{q\} = \{Q\}$$

Or

$$[k]\{q\} = \{Q\} \tag{6.5}$$

where $[k]$ is the element stiffness matrix given by

$$[k] = \int [B]^{\mathrm{T}} [D] [B] \mathrm{d}V. \tag{6.6}$$

The stiffness matrix given by Eq. (6.6) will be exact if the displacement shape function assumed in Eq. (6.2) is exact.

6.5 St. Venant Torsion Element

The St. Venant torsion element is analogous to axial displacement finite element since axially loaded and St. Venant torsion problems are similar. The finite element shown in Fig. 6.2 has two generalized coordinates (i.e., two degrees of freedom), one at each end. These are the torsional rotations at the ends. The torsional rotation, ϕ, at any point distant z from the left end is assumed to be linearly related to the degrees of freedom ϕ_1 and ϕ_2. Hence, with $\xi = \frac{z}{L}$

$$\phi = (1 - \xi)\phi_1 + \xi\phi_2. \tag{6.7}$$

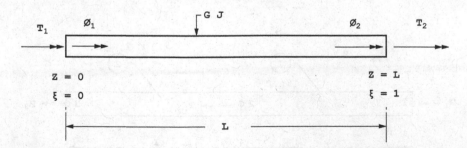

Fig. 6.2 St. Venant torsion element

The linear relationship is exact for St. Venant torsion problems loaded with point torques. Hence, the stiffness matrix that will be obtained by using this shape function will be exact. We have

$$\phi' = \frac{d\phi}{dz} = \frac{1}{L}[-1\ 1]\begin{Bmatrix} \phi_1 \\ \phi_2 \end{Bmatrix}$$

Thus,

$$[B] = \frac{1}{L}[-1\ 1]$$

Hence,

$$[k] = \int_0^1 \frac{1}{L}\begin{Bmatrix} -1 \\ 1 \end{Bmatrix} GJ\frac{1}{L}[-1\ 1]Ld\xi$$

$$= \frac{GJ}{L}\begin{bmatrix} 1 & -1 \\ -1 & 1 \end{bmatrix} \tag{6.8}$$

6.6   Applications of St. Venant Torsion Element

As an application of the St. Venant torsion element shown in Fig. 6.2, the structure shown in Fig. 5.1 is considered. It is modeled with two St. Venant torsion finite elements as shown in Fig. 6.3. The structure has three nodes. The finite element analysis proceeds as follows:

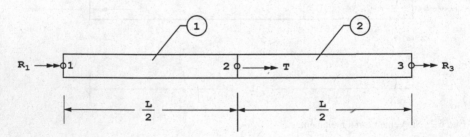

Fig. 6.3 Two St. Venant torsion elements modeling the structure

[a]   element stiffness matrix in structure coordinates

$$\text{element (1)} \quad [k] = \frac{2GJ}{L} \begin{array}{cc} \phi_1 & \phi_2 \end{array} \begin{bmatrix} 1 & -1 \\ -1 & 1 \end{bmatrix}$$

$$\text{element (2)} \quad [k] = \frac{2GJ}{L} \begin{array}{cc} \phi_2 & \phi_3 \end{array} \begin{bmatrix} 1 & -1 \\ -1 & 1 \end{bmatrix}$$

[b]   assembling the element stiffnesses to form the structure stiffness

$$[K] = \frac{2GJ}{L} \begin{array}{ccc} \phi_1 & \phi_2 & \phi_3 \end{array} \begin{bmatrix} 1 & -1 & 0 \\ -1 & 2 & -1 \\ 0 & -1 & 1 \end{bmatrix}$$

Thus, the force–deflection characteristics of the structure is

$$\frac{2GJ}{L} \begin{bmatrix} 1 & -1 & 0 \\ -1 & 2 & -1 \\ 0 & -1 & 1 \end{bmatrix} \begin{Bmatrix} \phi_1 \\ \phi_2 \\ \phi_3 \end{Bmatrix} = \begin{Bmatrix} R_1 \\ T \\ R_3 \end{Bmatrix}$$

[c]   application of the boundary conditions to get the restrained structure:

We have $\phi_1 = \phi_3 = 0$. Hence, deleting the first row and the first column as well as the third row and the third column, we obtain

$$\frac{2GJ}{L}[2]\phi_2 = T$$

which gives

$$\phi_2 = \frac{TL}{4GJ}$$

[d]   determination of the element torques

$$\text{element (1)} \quad \begin{Bmatrix} T_1 \\ T_2 \end{Bmatrix} = \frac{2GJ}{L} \begin{bmatrix} 1 & -1 \\ -1 & 1 \end{bmatrix} \begin{Bmatrix} 0 \\ \frac{TL}{4GJ} \end{Bmatrix} = \begin{Bmatrix} -T/2 \\ T/2 \end{Bmatrix}$$

$$\text{element (2)} \quad \begin{Bmatrix} T_1 \\ T_2 \end{Bmatrix} = \frac{2GJ}{L} \begin{bmatrix} 1 & -1 \\ -1 & 1 \end{bmatrix} \begin{Bmatrix} \frac{TL}{4GJ} \\ 0 \end{Bmatrix} = \begin{Bmatrix} T/2 \\ -T/2 \end{Bmatrix}$$

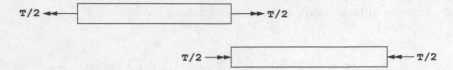

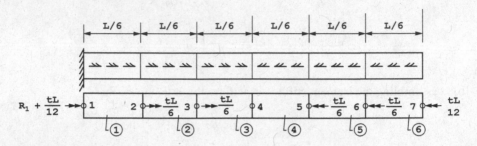

Fig. 6.4 Element torque

Fig. 6.5 Six element model

The element torques are shown in Fig. 6.4. The next example is the structure shown in Fig. 5.2. Here the torsional rotation varies quadratically with respect to z, and hence, the linear St. Venant torsional element will not be able to model the problem exactly. Thus, a large number of linear finite elements must be used, and the non-generalized distributed torques must be lumped as concentrated torques at the generalized coordinates. The finite element modeling of the problem with 6 St. Venant torsional finite elements is shown in Fig. 6.5.

The stiffness matrix of each of the six finite elements is of the form

$$[k] = \frac{6GJ}{L} \begin{bmatrix} 1 & -1 \\ -1 & 1 \end{bmatrix}$$

Assembling the element stiffness matrices, we get

$$\frac{6GJ}{L} \begin{bmatrix} 2 & -1 & & & & & \\ -1 & 2 & -1 & & & & \\ & -1 & 2 & -1 & & & \\ & & -1 & 2 & -1 & & \\ & & & -1 & 2 & -1 & \\ & & & & -1 & 2 & -1 \\ & & & & & -1 & 1 \end{bmatrix} \begin{Bmatrix} \phi_1 \\ \phi_2 \\ \phi_3 \\ \phi_4 \\ \phi_5 \\ \phi_6 \\ \phi_7 \end{Bmatrix} = \begin{Bmatrix} R_1 + tL/12 \\ tL/6 \\ tL/6 \\ 0 \\ -tL/6 \\ -tL/6 \\ -tL/12 \end{Bmatrix}$$

Applying the boundary condition $\phi_1 = 0$, we get

$$\frac{6GJ}{L}\begin{bmatrix} 2 & -1 & & & & \\ -1 & 2 & -1 & & & \\ & -1 & 2 & -1 & & \\ & & -1 & 2 & -1 & \\ & & & -1 & 2 & -1 \\ & & & & -1 & 1 \end{bmatrix}\begin{Bmatrix} \phi_2 \\ \phi_3 \\ \phi_4 \\ \phi_5 \\ \phi_6 \\ \phi_7 \end{Bmatrix} = \frac{tL}{12}\begin{Bmatrix} 2 \\ 2 \\ 0 \\ -2 \\ -2 \\ -1 \end{Bmatrix}$$

Solving we get

$$\begin{Bmatrix} \phi_2 \\ \phi_3 \\ \phi_4 \\ \phi_5 \\ \phi_6 \\ \phi_7 \end{Bmatrix} = \frac{tL^2}{72GJ}\begin{Bmatrix} -1 \\ -4 \\ -9 \\ -14 \\ -17 \\ -18 \end{Bmatrix}$$

The maximum rotation is $\frac{tL^2}{4GJ}$. This as well as the rotations at other nodes coincide with the exact values. However, in between the nodes, the finite element analysis with the linear St. Venant torsion element gives a linear variation of rotation while the actual variation of rotation is parabolic. The distribution of element torques given by the finite element analysis is step-wise constant while the actual variation is linear. These are shown in Fig. 6.6. It is seen that the finite element analysis with 6 elements underestimates the torque by as much as 16.7%.

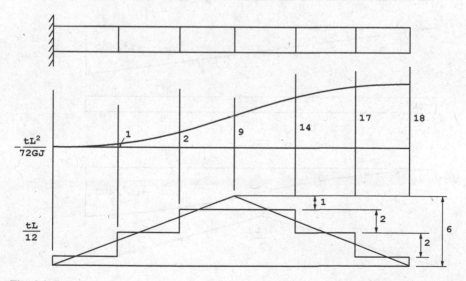

Fig. 6.6 Rotation and toque given by the finite element analysis

An exact analysis using the linear element is, however, possible by analyzing the problem in two stages. In stage-I, the structure is held fixed at the right and (by deploying the person with the jack whom we used before!). In stage-II, the person must be removed (after paying him well for his service!) which can be simulated by applying the reversed fixed end forces as shown in Fig. 6.7. The stage-II analysis consists only of generalized forces and hence can be carried out by modeling the structure with just one St. Venant linear finite element of Fig. 6.2. That is,

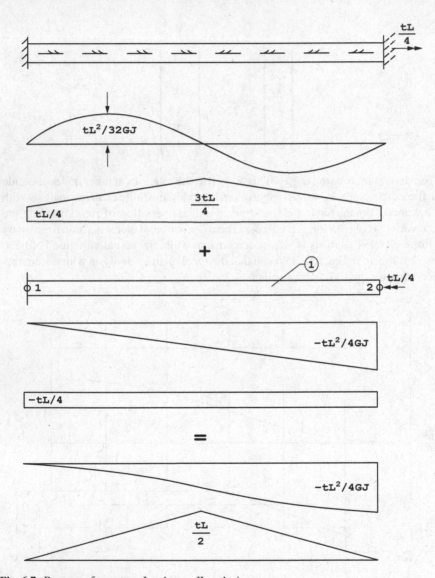

Fig. 6.7 Response from stage-I and stage-II analysis

$$\frac{GJ}{L}\begin{bmatrix} 1 & -1 \\ -1 & 1 \end{bmatrix}\begin{Bmatrix} \phi_1 \\ \phi_2 \end{Bmatrix} = \begin{Bmatrix} R_1 \\ -tL/4 \end{Bmatrix}$$

Applying the boundary condition $\phi_1 = 0$, we get

$$\frac{GJ}{L}[1]\phi_2 = -\frac{tL}{4}$$

$$\phi_2 = -\frac{tL^2}{4GJ}$$

The rotation and torque distributions for the stage-I and stage-II analyses are shown in Fig. 6.7. The response of the structure (torque or rotation) is obtained by adding the corresponding responses in stages I and II. The response is shown in Fig. 6.7. It can be seen that the response is exact, and no approximations are involved anywhere.

6.7 Vlasov Torsion Element

Pure warping torsion problems can be analyzed using the Vlasov torsion element as shown in Fig. 6.8. Basically it is derived from a Bernoulli–Euler beam finite element since the warping torsion and flexural problems are analogous. The stiffness matrix of the finite element is

$$[k] = \frac{2EI_\Omega}{L^3}\begin{bmatrix} 6 & 3L & -6 & 3L \\ 3L & 2L^2 & -3L & L^2 \\ -6 & -3L & 6 & -3L \\ 3L & L^2 & -3L & 2L^2 \end{bmatrix} \tag{6.9}$$

As an application of the finite element, the structure of Fig. 5.1 is modeled in Fig. 6.9 with two Vlasov torsion finite elements. The stiffness matrix of each of these finite elements is

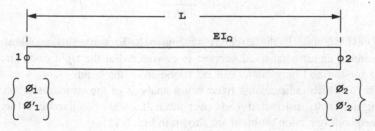

Fig. 6.8 Vlasov torsion element

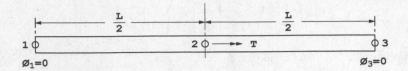

Fig. 6.9 Finite element model of the structure shown in Fig. 5.1

$$[k] = \frac{4EI_\Omega}{L^3} \begin{bmatrix} 24 & 6L & -24 & 6L \\ 6L & 2L^2 & -6L & L^2 \\ -24 & -6L & 24 & -6L \\ 6L & L^2 & -6L & 2L^2 \end{bmatrix}$$

Assembling the stiffness matrices, we get

$$\frac{4EI_\Omega}{L^3} \begin{bmatrix} 24 & 6L & -24 & 6L & 0 & 0 \\ 6L & 2L^2 & -6L & L^2 & 0 & 0 \\ -24 & -6L & 48 & 0 & -24 & 6L \\ 6L & L^2 & 0 & 4L^2 & -6L & L^2 \\ 0 & 0 & -24 & -6L & 24 & -6L \\ 0 & 0 & 6L & L^2 & -6L & 2L^2 \end{bmatrix} \begin{Bmatrix} \phi_1 \\ \phi_1' \\ \phi_2 \\ \phi_2' \\ \phi_3 \\ \phi_3' \end{Bmatrix} = \begin{Bmatrix} R_1 \\ 0 \\ T \\ 0 \\ R_5 \\ 0 \end{Bmatrix}$$

In the foregoing, R_1 and R_5 are the torsion offered by the left and right supports, respectively. Applying the boundary conditions $\phi_1 = \phi_3 = 0$, corresponding to the end supports, we get

$$\frac{4EI_\Omega}{L^3} \begin{bmatrix} 2L^2 & -6L & L^2 & 0 \\ -6L & 48 & 0 & 6L \\ L^2 & 0 & 4L^2 & L^2 \\ 0 & 6L & L^2 & 2L^2 \end{bmatrix} \begin{Bmatrix} \phi_1' \\ \phi_2 \\ \phi_2' \\ \phi_3' \end{Bmatrix} = \begin{Bmatrix} 0 \\ T \\ 0 \\ 0 \end{Bmatrix}$$

Solving we get

$$\phi_2 = \frac{TL^3}{48EI_\Omega}$$

The next example is the structure of Fig. 5.2. To solve this problem in an exact manner, an artificial fixed support is considered at the right end. The fixed warping torsion and bimoments can be found from the beam analogy as shown in Fig. 6.10, which indicates the fixed beam analysis of the structure subjected to opposing uniformly distributed loads over the half spans. The fixed end forces for the corresponding torsion problem are shown in Fig. 6.11.

The finite elements analysis of stage-II is conducted on the structure by applying reversed fixed end forces as shown in Fig. 6.11. We have

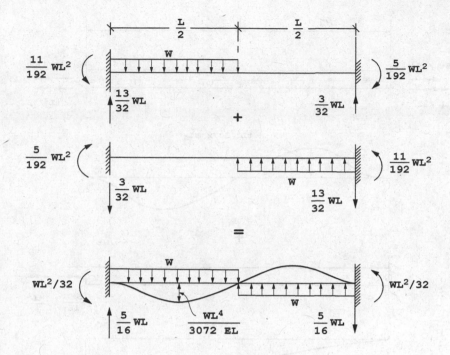

Fig. 6.10 Fixed warping torsion and bimoments from the beam analogy

$$\frac{2EI_\Omega}{L^3}\begin{bmatrix} 6 & 3L & -6 & 3L \\ 3L & 2L^2 & -3L & L^2 \\ -6 & -3L & 6 & -3L \\ 3L & L^2 & -3L & 2L^2 \end{bmatrix}\begin{Bmatrix} \phi_1 \\ \phi_1' \\ \phi_2 \\ \phi_2' \end{Bmatrix}=\begin{Bmatrix} R_1 \\ R_2 \\ -5tL/16 \\ tL^2/32 \end{Bmatrix}$$

Applying the boundary conditions $\phi_1 = \phi_1' = 0$, we get

$$\frac{2EI_\Omega}{L^3}\begin{bmatrix} 6 & -3L \\ -3L & 2L^2 \end{bmatrix}\begin{Bmatrix} \phi_2 \\ \phi_2' \end{Bmatrix}=-\frac{tL^2}{32}\begin{Bmatrix} 10/L \\ -1 \end{Bmatrix}$$

Solving we have

$$\begin{Bmatrix} \phi_2 \\ \phi_2' \end{Bmatrix}=\frac{tL^3}{192EI_\Omega}\begin{Bmatrix} -17L \\ -24 \end{Bmatrix}$$

6.8 Mixed Torsion Finite Element

The mixed torsion problem is governed by the relation

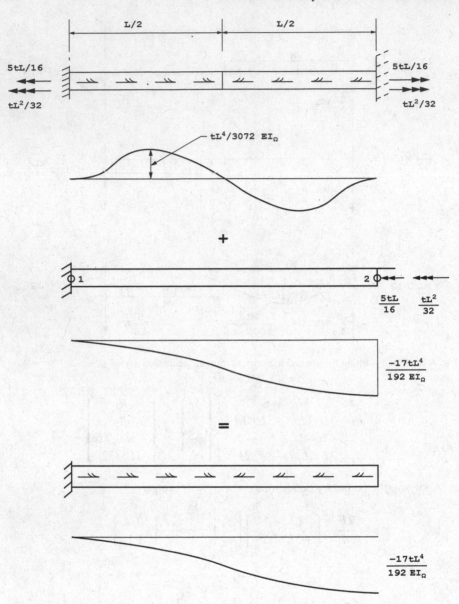

Fig. 6.11 Fixed end forces

$$EI_\Omega \phi'''' - GJ\phi'' = m_z \tag{6.10}$$

The behavior of a beam subjected to a tensile axial load P is governed by the relation

$$EIV'''' - PV'' = q \tag{6.11}$$

Equations (6.10) and (6.11) are analogous. The following are the analogous quantities.

Beam bending with tensile axial load	Mixed torsion
Deflection, V	Rotation, ϕ
Slope, V'	Warping, ϕ'
Bending moment	Bimoment
Shear force	Twisting moment
Point load	Point torque
Distributed load	Distributed torque
P	GJ
I	I_Ω

The stiffness matrix of a mixed torsion finite element shown in Fig. 6.12 can therefore be written from the stiffness matrix of a beam element subjected to an axial load which is written in an exact manner in terms of the well-known stability functions. Thus, the stiffness matrix of the mixed torsion element is

$$[k] = \frac{EI_\Omega}{L^3} \begin{bmatrix} \alpha_1 & L\alpha_2 & -\alpha_1 & L\alpha_2 \\ L\alpha_2 & L^2\alpha_4 & -L\alpha_2 & L^2\alpha_3 \\ -\alpha_1 & -L\alpha_2 & \alpha_1 & -L\alpha_2 \\ L\alpha_2 & L^2\alpha_3 & -L\alpha_2 & L^2\alpha_4 \end{bmatrix} \tag{6.12}$$

where α_1 through α_4 are given by

$$\alpha_1 = 2s(1+c) + \pi^2\rho$$

$$\alpha_2 = s(1+c)$$

$$\alpha_3 = sc$$

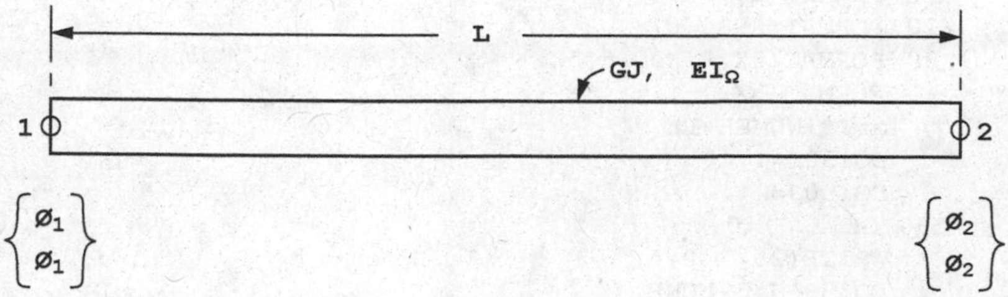

Fig. 6.12 Mixed torsion finite element

$$\alpha_4 = s$$

In which

$$\rho = \frac{GJL^2}{\pi^2 EI_\Omega}$$

$$\beta = \frac{\pi}{2}\sqrt{\rho}$$

$$s = \beta\left[\frac{1 - 2\beta\coth 2\beta}{\tanh\beta - \beta}\right]$$

And

$$c = \frac{2\beta - \sinh 2\beta}{\sinh 2\beta - 2\beta\cosh 2\beta}$$

The stiffness matrix given by Eq. 6.12 relates

$$\begin{Bmatrix} T_1 \\ M_{\Omega 1} \\ T_2 \\ M_{\Omega 2} \end{Bmatrix} = [k]\begin{Bmatrix} \phi_1 \\ \phi_1' \\ \phi_2 \\ \phi_2' \end{Bmatrix} \tag{6.13}$$

A computer program shown in the following does the torsional analysis of a member with arbitrary loads, multiple supports, and varying torsional properties.

```
    C234567890123456789012345678901234567890123456789
    C        1         2         3         4         5
    IMPLICIT REAL*8(A-H,D-Z)
    DIMENSION GK(40,40),EK(4,4),C(40,40),D(40),R(40),
    $NDOF(10)
    DIMENSION GAM(40),EI(40),GJ(40),AL(40),FACT(40),
    $X(4),F(4)
    READ(25,10)NUMEL
 10 FORMAT(I5)
 20 FORMAT(8F10.0)
 30 FORMAT(5X,4(D20.6,5X))
 40 FORMAT(5X,'        ')
    M=2*(NUMEL+1)
    DO150I=1,40
    DO150J=1,40
150 GK(I,J)=0.0
    WRITE(26, 40)
    DO350 I=1,NUMEL
```

```
          READ(25,20)GAM(I),EI(I),GJ(I),AL(I),FACT(I)
          WRITE(26,30)GAM(I),EI(I),GJ(I),AL(I),FACT(I)
 350 CONTINUE
          WRITE(26,40)
          DO100 K=1,NUMEL
          CALLSTIF(FACT(K),GAM(K),EI(K),GJ(K),AL(K),EK)
          DO200I=1,4
          I1=2*k-2+I
          DO200 J=1,4
          J1=2*k-2+J
 200 GK(I1,J1)=GK(I1,J1)+EK(I,J)
 100 CONTINUE
          READ (25,35)NUMR,(NDOF(I),I=1,NUMR)
  35 FORMAT(11I5)
          DO105 I=1,NUMR
          L=NDOF(I)
          DO106 J=1,M
          GK(L,J)=0.0
 106 GK(J,L)=0.0
          GK(L,L)=1.0
 105 CONTINUE
          DO107 I=1,M
 107 R(I)=0.0
          READ(25,35)NUML
          DO108 I=1,NUML
          READ(25,45)NLOAD,ALOAD
 108 R(NLOAD)=ALOAD
  45 FORMAT(I5,F10.0)
          WRITE(26,40)
          CALLINVERS(GK,C,M)
          DO400 I=1,M
          D(I)=0.0
          DO400 J=1,M
 400 D(I)=D(I)+GK(I,J)*R(J)
          WRITE(26,40)
          WRITE(26,30)(D(I),I=1,M)
          WRITE(26,40)
          DO500 I=1,NUMEL
          K1=2*I-1
          CALLSTIF(FACT(I),GAM(I),EI(I),GJ(I), AL(I),EK)
          DO510 K=1,4
          KK=K1-1+K
 510 X(K)=D(KK)
          DO520 K=1,4
          F(K)=0.0
          DO530 KK=1,4
 530 F(K)=F(K)+EK(K,KK)*X(KK)
 520 CONTINUE
          WRITE(26,30)(F(K),K=1,4)
          WRITE(26,40)
 500 CONTINUE
```

```
      WRITE(26,40)
      STOP
      END
      SUBROUTINE STIF(FACT,GAM,EI,GJ,AL,EK)
      IMPLICIT REAL*8(A-H,D-Z)
      DIMENSIONEK(4,4)
      PI=3.14159265359
      G=.5*AL*DSQRT(FACT*GJ/EI)
      R=FACT*GJ*(AL**2)/(PI*PI*EI)
      S1=DSINH(2.0*G)
      S2=DCOSH(2.0*G)
      S3=S1/S2
      S4=S2/S1
      S5=DSINH(G)/DCOSH(G)
      S=G*(1.0-2.0*G*S4)/(S5-G)
      C=(2.0*G-S1)/(S1-2.0*G*S2)
      U1=2.0*S*(1.0+C)+PI*PI*R
      U2=S*(1.0+C)
      U3=S*C
      U4=S
      C1=EI/(AL**3)
      C2=EI/(AL**2)
      C3=EI/AL
      EK(1,1)=C1*U1
      EK(1,2)=C2*U2
      EK(1,3)=-C1*U1
      EK(1,4)=C2*U2
      EK(2,2)=C3*U4
      EK(2,3)=-C2*U2
      EK(2,4)=C3*U3
      EK(3,3)=C1*U1
      EK(3,4)=-C2*U2
      EK(4,4)=C3*U4
      DO100 I=1,4
      DO100 J=1,4
  100 EK(J,I)=EK(I,J)
      DO200 I=1,4
  200 EK(I,2)=EK(I,2)/GAM
      RETURN
      END
      SUBROUTINE INVERS(A,U,N)
      IMPLICIT REAL*8 (A-H, O-Z)
      DIMENSION A(40,40),U(40,40)
      DO1I=1,N
      DO1J=1,N
      U(I,J)=0.0
      IF(I.EQ .J)U(I,J)=1.0
    1 CONTINUE
      EPS=1.0000001
      DO15 I=1,N
      K=I
```

```
      IF(I-N)21,7,21
 21   IF(A(I,I)-EPS)5,6,7
  5   IF(-A(I,I)-EPS)6,6,7
  6   K=K+1
      DO23 J=1,N
      U(I,J)=U(I,J)+U(K,J)
 23   A(1,J)=A(I,J)+A(K,J)
      GOTO21
  7   DIV=A(I,I)
      DO9J=1,N
      U(I,J)=U(I,J)/DIV
  9   A(I,J)=A(I,J)/DIV
      DO15 M=1,N
      DELT=A(M,I)
      IF(DABS(DELT)-EPS)15,15,16
 16   IF(M-I)10,15,10
 10   DO11 J=1,N
      U(M,J)=U(M,J)-U(I,J)*DELT
 11   A(M,J)=A(M,J)-A(I,J)*DELT
 15   CONTINUE
      DO34I=1,N
      DO34J=1,N
 34   A(I,J)=U(I,J)
      RETURN
      END
```

The program uses the mixed torsion finite element defined by Eq. (6.12). The program is also written in such a manner that it can take care of the warping compatibility at the junctions of finite elements of dissimilar cross sections. The warping compatibility factor γ is introduced into the stiffness matrix as follows.

$$\begin{Bmatrix} T_1 \\ M_{\Omega 1} \\ T_2 \\ M_{\Omega 2} \end{Bmatrix} = \begin{bmatrix} k_{11} & k_{12}/\gamma & k_{13} & k_{14} \\ k_{21} & k_{22}/\gamma & k_{23} & k_{24} \\ k_{31} & k_{32}/\gamma & k_{33} & k_{34} \\ k_{41} & k_{42}/\gamma & k_{43} & k_{44} \end{bmatrix} \begin{Bmatrix} \phi_1 \\ \phi_1' \\ \phi_2 \\ \phi_2' \end{Bmatrix} \tag{6.14}$$

In which the left hand node of the finite element has the stress resultants defined, for example, as

$$T_1 = \begin{bmatrix} k_{11} & k_{12} & k_{13} & k_{14} \end{bmatrix} \begin{Bmatrix} \phi_1 \\ \phi_1'/\gamma \\ \phi_2 \\ \phi_2' \end{Bmatrix}$$

$$= \begin{bmatrix} k_{11} & \frac{k_{12}}{\gamma} & k_{13} & k_{14} \end{bmatrix} \begin{Bmatrix} \phi_1 \\ \phi_1' \\ \phi_2 \\ \phi_2' \end{Bmatrix}$$

Thus, the generalized coordinates at the left end of a junction are taken as the global coordinates. The node at the left-hand side of a finite element has the right end of a junction, and hence, its ϕ' is taken as the $\frac{1}{\gamma}$ times the ϕ' which is at the left end of the junction. The stiffness matrix of the mixed torsion finite element is thus computed in the program taking into account the warping compatibility factor. It must be noted that the stiffness matrix is now unsymmetric because of the definition. The typical input data for the torsional analysis of a one-dimensional structure using this program is explained as follows:

1. First Card (data line) (Format I5)

    1–5 Number of finite elements used

2. Second Card (Set) (one for each finite element) (Format 8F10.0)

    1–10 Warping compatibility factor
    11–20 The value of EI_Ω
    21–30 The value of GJ
    31–40 The value of length

3. Third Card (Format 11 I5)

    1–5 The number of restrained degrees of freedom
    6–10 The degree of freedom of the first restraint
    11–20 The degree of freedom of the second restraint.
    And so on.

4. Fourth Card (Format I5)

    1–5 Number of applied loads

5. Fifth Card (Set) (Format I5, F10.0)

    1–5 The loaded degree of freedom for the first load
    6–15 The value of the first load.
    and so on with one card for each load.

The output from the program will consist of the ϕ and ϕ' values at all nodes as well as the torsion and bimoments at the end points of all finite elements.

As an example of finite element analysis, the structure of review problem 3 of the previous chapter is considered. The structure is modeled with two mixed torsion finite elements as shown in Fig. 6.13. The input data is as follows:

1. First Card

    1–5 2

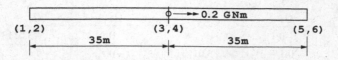

Fig. 6.13 Finite element model with two mixed torsion finite elements

2. Second Card Set

   First card in this set
   1–10 1.0
   11–20 2450000.0
   21–30 1006.0
   31–40 35.0
   Second card in this set.
   (same as the first card of this set).

3. Third Card:

   1–5 2
   6–10 1
   11–15 5

4. Fourth Card:

   1–5 1

5. Fifth Card:

   1–5 3
   6–15 0.2

The output given by the program is as follows

Degree of freedom	Generalized displacement
1	0.0
2	0.2065×10^{-4}
3	0.4858×10^{-3}
4	0
5	0
6	-0.2065×10^{-4}

Element	Q_1	Q_2	Q_3	Q_4
1	−0.1	0	0.1	−3.0113
2	0.1	3.0113	−0.1	0

The rotation at node 2 compares with the value obtained in the previous chapter based on the solution of the differential equation. The finite element analysis has yielded an exact solution because the stiffness matrix given by Eq. 6.12 is exact, and there are only nodal loads in the problem.

Further examples of torsional analysis using the finite element method are given in the review problems of this chapter. Applications of the finite element method to torsional instability problems are described in the next chapter.

The foregoing finite element analysis is based on the assumption that the cross section does not change its shape. A one-dimensional analysis, by modeling the structure only along the length, is therefore possible in this situation. This assumption must be realized by the provision of adequate diaphragms. However, the torsional analysis of structures can be performed similar to the analysis for any other form of loading through a three-dimensional modeling of the structure using thin-plate bending finite elements.

6.9 Finite Element Method for Warping Properties of Thin-Walled Sections

The warping properties such as normalized unit warping, ω_n, at various points on a thin-walled section can be easily determined by a finite element analysis. The unit warping is defined over a finite element using linear shape functions as follows:

$$\omega_n = (1 - \xi)\omega_{n1} + \xi\omega_{n2}$$

Using this function for ω_n in the variational problem of torsion, we get the 'stiffness matrix'

$$[k] = \frac{Gt}{L}\begin{bmatrix} 1 & -1 \\ -1 & 1 \end{bmatrix} \tag{6.15}$$

And the 'load vector'

$$Gth_{sc}\begin{Bmatrix} -1 \\ 1 \end{Bmatrix} \tag{6.16}$$

in which G is the rigidity modulus, t the thickness, and L the length of the finite element. The tangential distance of the finite element from the shear center is denoted by h_{sc}.

The given cross section is divided into a number of finite elements. The stiffness matrices and load vectors of these elements can be written and assembled in the usual manner. Each node has only one unknown, viz. the normalized unit warping, ω_n. At some of the nodes, e.g., nodes on the lines of symmetry, the normalized unit warping is zero. These are the 'boundary conditions' of the finite element problem. After applying the boundary conditions, the normalized unit warpings at the remaining nodes can be solved.

As an example, the cross section shown in Fig. 6.14 is considered. The finite element modeling of one-half of the section is also given in the same figure.

The element stiffness matrices and load vectors of the various finite elements are written using Eqs. (6.15) and (6.16). The common terms Gt have been dropped from these expressions. In writing the load vector, a clockwise sweep of the element about the shear center is taken as positive. All elements (1–2, 2–4, 3–4, 4–6, and 6–5) are taken with a counterclockwise aspect with respect to the shear center for illustration in the following (if we take element 3 for example as 4–3 instead of 3–4, it would have a clockwise aspect).

$$\text{element 1:} \quad \frac{2}{a}\begin{bmatrix} 1 & -1 \\ -1 & 1 \end{bmatrix}\begin{Bmatrix} \omega_{n1} \\ \omega_{n2} \end{Bmatrix} = 1.39a\begin{Bmatrix} 1 \\ -1 \end{Bmatrix}$$

$$\text{element 2:} \quad \frac{1}{a}\begin{bmatrix} 1 & -1 \\ -1 & 1 \end{bmatrix}\begin{Bmatrix} \omega_{n2} \\ \omega_{n4} \end{Bmatrix} = 0.5a\begin{Bmatrix} 1 \\ -1 \end{Bmatrix}$$

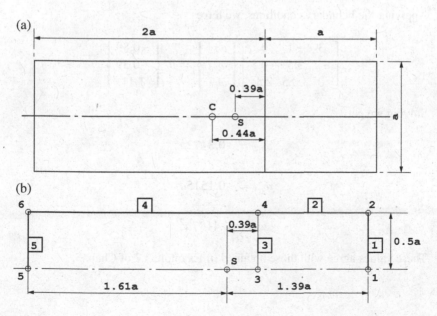

Fig. 6.14 **a** A two-cell cross section. **b** Finite element model

element 3: $\dfrac{2}{a}\begin{bmatrix} 1 & -1 \\ -1 & 1 \end{bmatrix}\begin{Bmatrix} \omega_{n3} \\ \omega_{n4} \end{Bmatrix} = 0.39a\begin{Bmatrix} 1 \\ -1 \end{Bmatrix}$

element 4: $\dfrac{1}{2a}\begin{bmatrix} 1 & -1 \\ -1 & 1 \end{bmatrix}\begin{Bmatrix} \omega_{n4} \\ \omega_{n6} \end{Bmatrix} = 0.5a\begin{Bmatrix} 1 \\ -1 \end{Bmatrix}$

element 5: $\dfrac{2}{a}\begin{bmatrix} 1 & -1 \\ -1 & 1 \end{bmatrix}\begin{Bmatrix} \omega_{n6} \\ \omega_{n5} \end{Bmatrix} = 1.61a\begin{Bmatrix} 1 \\ -1 \end{Bmatrix}$

Assembling the element stiffnesses and load vectors, we have

$$\frac{1}{a}\begin{bmatrix} 2 & -2 & 0 & 0 & 0 & 0 \\ -2 & 3 & 0 & -1 & 0 & 0 \\ 0 & 0 & 2 & -2 & 0 & 0 \\ 0 & -1 & -2 & 3.5 & 0 & -0.5 \\ 0 & 0 & 0 & 0 & 2 & -2 \\ 0 & 0 & 0 & -0.5 & -2 & 2.5 \end{bmatrix}\begin{Bmatrix} \omega_{n1} \\ \omega_{n2} \\ \omega_{n3} \\ \omega_{n4} \\ \omega_{n5} \\ \omega_{n6} \end{Bmatrix} = a\begin{Bmatrix} 1.39 \\ -0.89 \\ 0.39 \\ -0.39 \\ -1.61 \\ 1.11 \end{Bmatrix}$$

The boundary conditions of the problem relate to the nodes lying on the axis of symmetry where the warping is zero.

Thus,

$$\omega_{n1} = \omega_{n3} = \omega_{n5} = 0$$

Applying the boundary conditions, we have

$$\begin{bmatrix} 3 & -1 & 0 \\ -1 & 3.5 & -0.5 \\ 0 & -0.5 & 2.5 \end{bmatrix}\begin{Bmatrix} \omega_{n2} \\ \omega_{n4} \\ \omega_{n6} \end{Bmatrix} = a^2\begin{Bmatrix} -0.89 \\ -0.39 \\ 1.11 \end{Bmatrix}$$

Solving we get

$$\omega_{n2} = -0.3472a^2$$

$$\omega_{n4} = -0.1515a^2$$

$$\omega_{n6} = 0.4137a^2$$

These values agree with those obtained in Example 3.7 of Chap. 3.

6.10 Review Problems

1. Determine the distribution of torsional rotation in the structure shown in Fig. 6.15 in St. Venant torsion. Also determine the torques at various points. Use the finite element method. For portion 1–2, $G = 80$ kN/mm^2 and $J = 1046 \times 10^4$ mm^4. For portion 2–3, $G = 27$ kN/mm^2 and $J = 331 \times 10^4$ mm^4.
2. Solve the structure shown in Fig. 5.2 and considered in review problem 4 of the previous chapter using mixed torsion finite elements. The right end is free to warp.
3. Solve the foregoing problem if the right end is restrained against warping.
4. Solve the structure shown in Fig. 5.3 and considered in review problem 6 of the previous chapter using mixed torsion finite elements. Neglect deformations due to warping shear.
5. Solve the structure shown in Fig. 6.16 using the finite element method. The fork supports prevent twisting but not warping. $E = 206$ GPa, $G = 82.4$ GPa, $I_\Omega = 34.133 \times 10^6$ mm^6 and $J = 426.7$ mm^4
6. Determine the normalized unit warping distribution in the channel section as shown in Fig. 6.17.
7. Determine the normalized unit warping distribution in the multicellular section shown in Fig. 6.18. The shear center is located at a distance of 5723 mm below the node 7.
8. Write a computer program to determine the normalized unit warping distribution using the finite element method. The input data can be assumed as the number of elements, number of nodes and for each element, the thickness, the length, the tangential distance from the shear center and the end nodes with a

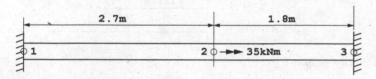

Fig. 6.15 Review problem 6.10.1

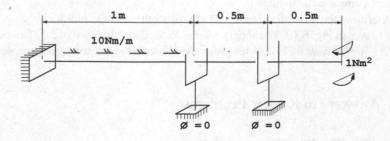

Fig. 6.16 Review problem 6.10.5

Fig. 6.17 Review
problem 6.10.6

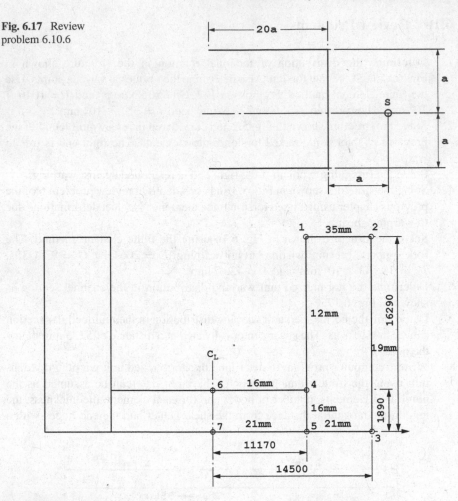

Fig. 6.18 Review problem 6.10.7

clockwise aspect about the shear center. The nodes with known zero warping
must complete the input.

9. Determine the normalized unit warping distribution in the multicellular section
    as shown in Fig. 6.19. The shear center is located at a distance of 5.526 m below
    the node 18. Use the computer program developed in the foregoing problem.

6.11 Answers to Review Problems

1. We have the stiffness matrix for element 1–2

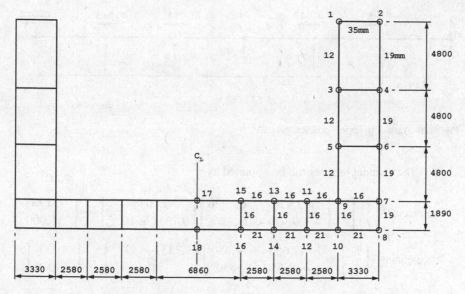

Fig. 6.19 Review problem 6.10.9

$$[k] = \frac{(0.8)(10^2)(1046 \times 10^4)}{2700} \begin{bmatrix} 1 & -1 \\ -1 & 1 \end{bmatrix}$$

$$= \begin{bmatrix} 309,926 & -309,926 \\ -309,926 & 309,926 \end{bmatrix}$$

similarly for the element 2–3,

$$[k] = \frac{(0.27)(10^2)(331 \times 10^4)}{1800} \begin{bmatrix} 1 & -1 \\ -1 & 1 \end{bmatrix}$$

$$= \begin{bmatrix} 49,650 & -49,650 \\ -49,650 & 49,650 \end{bmatrix}$$

Assembling the stiffness matrices, we have

$$\begin{bmatrix} 309,926 & -309,926 & 0 \\ -309,926 & 359,576 & -49,650 \\ 0 & -49,650 & 49,650 \end{bmatrix} \begin{Bmatrix} \phi_1 \\ \phi_2 \\ \phi_3 \end{Bmatrix} = \begin{Bmatrix} T_{R1} \\ 35,000 \\ T_{R2} \end{Bmatrix}$$

Applying the boundary conditions $\phi_1 = \phi_3 = 0$, we get

$$\phi_2 = \frac{35,000}{359,576} = 9.7337 \times 10^{-2} \text{ radian.}$$

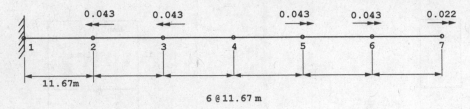

Fig. 6.20 Answer to review problem 6.10.2

The element torques can be obtained as

$$\text{element } 1 \begin{Bmatrix} T_1 \\ T_2 \end{Bmatrix} = \begin{bmatrix} 309{,}926 & -309{,}926 \\ -309{,}926 & 309{,}926 \end{bmatrix} \begin{Bmatrix} 0 \\ 9.7337 \times 10^{-2} \end{Bmatrix} = \begin{Bmatrix} -31{,}000 \\ 31{,}000 \end{Bmatrix}$$

$$\text{element } 2 \begin{Bmatrix} T_1 \\ T_2 \end{Bmatrix} = \begin{bmatrix} 49{,}650 & -49{,}650 \\ -49{,}650 & 49{,}650 \end{bmatrix} \begin{Bmatrix} 9.7337 \times 10^{-2} \\ 0 \end{Bmatrix} = \begin{Bmatrix} 4000 \\ -4000 \end{Bmatrix}$$

2.   The problem is solved by discretising the structure with six finite elements as shown in Fig. 6.20. The input data with this discretization is as follows:

Card No	Format	Data entered on card
1	I5	6
2	5F10.0	1.0, 2450000.0, 1006.0, 11.67
3 to 7	5F10.0	1.0, 2450000.0, 1006.0, 11.67
8	3 I5	2, 1, 2
9	I5	5
10	I5, F10.0	3, −0.0433
11	I5, F10.0	5, −0.0433
12	I5, F10.0	9, 0.0433
13	I5, F10.0	11, 0.0433
14	I5, F10.0	13, 0.0217

The output from the program is as follows:

Node	$\phi \times 10^4$	$\phi' \times 10^4$
1	0	0
2	0.7444	0.1262
3	2.8783	0.2353
4	6.0982	0.3092
5	9.9264	0.3401
6	13.9195	0.3416

(continued)

(continued)

Node	$\phi \times 10^4$	$\phi' \times 10^4$
7	17.8783	0.3382

3. The input data is the same as in the previous example with the following change in card no 8

Card No	Format	Data entered on card
8	4I5	3, 1, 2, 14

The output from the program is as follows:

Node	$\phi \times 10^4$	$\phi' \times 10^4$
1	0	0
2	0.5034	0.0847
3	1.9008	0.1500
4	3.8474	0.1753
5	5.7940	0.1500
6	7.1914	0.0847
7	7.6948	0

4. The problem is solved with 4 finite elements. The output is as follows:

Node	$\phi \times 10^2$	$\phi' \times 10^2$
1	0	0.3949
2	0.1108	0.3186
3	0.1759	0.0894
4	0.1932	0.0197
5	0.1991	0.0193

5. The leftmost span is loaded with a uniformly distributed torque. Hence, the problem can be solved in two stages as indicated in the text. It can also be solved by taking a large number of finite elements in the uniformly loaded span. The discretization using four finite elements in this span and the associated lumped loading are shown in Fig. 6.21. The input data with this discretization is as follows:

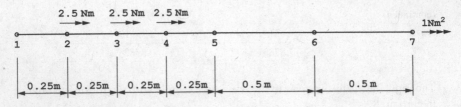

Fig. 6.21 Answer to review problem 6.10.5

Card No	Format	Data entered on card
1	I5	6
2	5F10.0	1.0, 7.031398, 35.16008, 0.25
3 to 5	5F10.0	1.0, 7.031398, 35.16008, 0.25
6	5F10.0	1.0, 7.031398, 35.16008, 0.5
7	5F10.0	1.0, 7.031398, 35.16008, 0.5
8	5 I5	4, 1, 2, 9, 11
9	I5	4
10	I5, F10.0	3, 2.5
11	I5, F10.0	5, 2.5
12	I5, F10.0	7, 2.5
13	I5, F10.0	14, 1.0

The output from the computer program is given in the following. The bimoment at the fixed support is seen to be 0.8732 Nm^2

Node	ϕ	ϕ'
1	0.0	0.0
2	0.00237	0.01313
3	0.00463	0.00277
4	0.00343	−0.01160
5	0	−0.01179
6	0	0.01285
7	0.01628	0.05891

Element	Torsion		Bimoment	
	Left end	Right end	Left end	Right end
1	−4.2873	4.2873	−0.8732	−0.1153
2	−1.7873	1.7873	0.1153	−0.4826
3	0.7127	−0.7127	0.4826	−0.3468

(continued)

(continued)

Element	Torsion		Bimoment	
	Left end	Right end	Left end	Right end
4	3.2128	−3.2128	0.3468	0.3360
5	0.1834	−0.1834	−0.3360	0.4277
6	0	0	−0.4277	1.0000

The problem has also been solved with 10 finite elements in the first span. The bimoment at the fixed support with this finer discretization is found to be 0.9183 Nm^2 (the exact value is 0.9269 Nm^2.

6. The finite element discretization of the channel (the upper half) is shown in Fig. 6.22. We have the following stiffness matrices and load vectors (the common term Gt is dropped from both).

$$\text{Element 1:} \frac{1}{2a}\begin{bmatrix} 1 & -1 \\ -1 & 1 \end{bmatrix}\begin{Bmatrix} \omega_{n1} \\ \omega_{n2} \end{Bmatrix} = a\begin{Bmatrix} -1 \\ 1 \end{Bmatrix}$$

$$\text{Element 2:} \frac{1}{a}\begin{bmatrix} 1 & -1 \\ -1 & 1 \end{bmatrix}\begin{Bmatrix} \omega_{n2} \\ \omega_{n3} \end{Bmatrix} = a\begin{Bmatrix} 1 \\ -1 \end{Bmatrix}$$

Assembling we have

$$\frac{1}{2a}\begin{bmatrix} 1 & -1 & 0 \\ -1 & 3 & -2 \\ 0 & -2 & 2 \end{bmatrix}\begin{Bmatrix} \omega_{n1} \\ \omega_{n2} \\ \omega_{n3} \end{Bmatrix} = a\begin{Bmatrix} -1 \\ 2 \\ -1 \end{Bmatrix}$$

Applying the boundary conditions $\omega_{n3} = 0$, we get

$$\begin{bmatrix} 1 & -1 \\ -1 & 3 \end{bmatrix}\begin{Bmatrix} \omega_{n1} \\ \omega_{n2} \end{Bmatrix} = 2a^2\begin{Bmatrix} -1 \\ 2 \end{Bmatrix}$$

Fig. 6.22 Answer to review problem 6.10.6

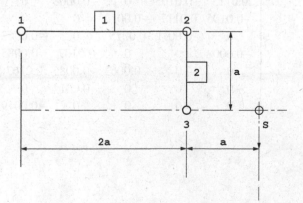

Solving we get

$$\omega_{n1} = -a^2$$

$$\omega_{n2} = a^2$$

7. The stiffness matrices and load vectors of the elements are (G is dropped from both)

$$\text{element 1: } \frac{0.035}{3.33}\begin{bmatrix} 1 & -1 \\ -1 & 1 \end{bmatrix}\begin{Bmatrix} \omega_{n1} \\ \omega_{n2} \end{Bmatrix} = (0.035)(22.013)\begin{Bmatrix} -1 \\ 1 \end{Bmatrix}$$

$$\text{element 2: } \frac{0.019}{16.29}\begin{bmatrix} 1 & -1 \\ -1 & 1 \end{bmatrix}\begin{Bmatrix} \omega_{n2} \\ \omega_{n3} \end{Bmatrix} = (0.019)(14.5)\begin{Bmatrix} -1 \\ 1 \end{Bmatrix}$$

$$\text{element 3: } \frac{0.012}{14.4}\begin{bmatrix} 1 & -1 \\ -1 & 1 \end{bmatrix}\begin{Bmatrix} \omega_{n1} \\ \omega_{n4} \end{Bmatrix} = (0.012)(11.17)\begin{Bmatrix} -1 \\ 1 \end{Bmatrix}$$

$$\text{element 4: } \frac{0.016}{1.89}\begin{bmatrix} 1 & -1 \\ -1 & 1 \end{bmatrix}\begin{Bmatrix} \omega_{n4} \\ \omega_{n5} \end{Bmatrix} = (0.016)(11.17)\begin{Bmatrix} -1 \\ 1 \end{Bmatrix}$$

$$\text{element 5: } \frac{0.016}{11.17}\begin{bmatrix} 1 & -1 \\ -1 & 1 \end{bmatrix}\begin{Bmatrix} \omega_{n6} \\ \omega_{n4} \end{Bmatrix} = (0.016)(7.613)\begin{Bmatrix} -1 \\ 1 \end{Bmatrix}$$

$$\text{element 6: } \frac{0.021}{11.17}\begin{bmatrix} 1 & -1 \\ -1 & 1 \end{bmatrix}\begin{Bmatrix} \omega_{n7} \\ \omega_{n5} \end{Bmatrix} = (0.021)(5.723)\begin{Bmatrix} -1 \\ 1 \end{Bmatrix}$$

$$\text{element 7: } \frac{0.021}{3.33}\begin{bmatrix} 1 & -1 \\ -1 & 1 \end{bmatrix}\begin{Bmatrix} \omega_{n5} \\ \omega_{n3} \end{Bmatrix} = (0.021)(5.723)\begin{Bmatrix} -1 \\ 1 \end{Bmatrix}$$

Assembling the stiffness matrices and load vectors, we have

$$\begin{bmatrix} 0.0113 & -0.0105 & 0 & -0.0008 & 0 & 0 & 0 \\ -0.0105 & 0.0117 & -0.0012 & 0 & 0 & 0 & 0 \\ 0 & -0.0012 & 0.0075 & 0 & -0.0063 & 0 & 0 \\ -0.0008 & 0 & 0 & 0.0107 & -0.0085 & -0.0014 & 0 \\ 0 & 0 & -0.0063 & -0.0085 & 0.0167 & 0 & -0.0019 \\ 0 & 0 & 0 & -0.0014 & 0 & 0.0014 & 0 \\ 0 & 0 & 0 & 0 & -0.0019 & 0 & 0.0019 \end{bmatrix}\begin{Bmatrix} \omega_{n1} \\ \omega_{n2} \\ \omega_{n3} \\ \omega_{n4} \\ \omega_{n5} \\ \omega_{n6} \\ \omega_{n7} \end{Bmatrix}$$

$$= \begin{Bmatrix} -0.9045 \\ 0.4950 \\ 0.3957 \\ 0.0771 \\ 0.1787 \\ -0.1218 \\ -0.1202 \end{Bmatrix}$$

Applying the boundary conditions $\omega_{n6} = \omega_{n7} = 0$, we have

$$\begin{bmatrix} 0.0113 & -0.0105 & 0 & -0.0008 & 0 \\ -0.0105 & 0.0117 & -0.0012 & 0 & 0 \\ 0 & -0.0012 & 0.0075 & 0 & -0.0063 \\ -0.0008 & 0 & 0 & 0.0107 & -0.0085 \\ 0 & 0 & -0.0063 & -0.0085 & 0.0167 \end{bmatrix} \begin{Bmatrix} \omega_{n1} \\ \omega_{n2} \\ \omega_{n3} \\ \omega_{n4} \\ \omega_{n5} \end{Bmatrix} = \begin{Bmatrix} -0.9045 \\ 0.4950 \\ 0.3957 \\ 0.0771 \\ 0.1787 \end{Bmatrix}$$

Solving we get

$$\omega_{n1} = -156.2\,\text{m}^2 \quad \omega_{n2} = -87.3\,\text{m}^2 \quad \omega_{n3} = 109.2\,\text{m}^2$$
$$\omega_{n4} = 60.3\,\text{m}^2 \ \text{ and } \ \omega_{n5} = 82.8\,\text{m}^2$$

8.   The computer program is shown in the following. The program reads the number of elements (NUMEL) and the number of nodes (NTOT). The global stiffness matrix (GK) is then assembled in the 350 loop. For each element, the thickness (T), the length (AL), and the tangential distance from the shear center (DIST) are read. Also read are the end nodes ($N1$ and $N2$) of the element given with a clockwise aspect with respect to the shear center ($N1$–$N2$ has a clockwise aspect with respect to the shear center). The global load vector (D) is also assembled in this loop.

The total number of nodes (NUMR) where warping is known to be zero is then read. Also read are the node numbers of each of the NUMR nodes. The boundary conditions are applied in the 105 loop. The inverse of GK is returned in GK itself which then premultiplies the D vector to give the vector of normalized unit warping at various nodes (R).

```
      C23456789012345678901234567890123456789012345678901234567890
      C         1         2         3         4         5
        IMPLICIT REAL *8 (A-H, D-Z)
        DIMENSION GK(20,20),D(20,20),D(20),R(20),  .
        $NDOF (10)
    10  FORMAT(2I5)
    20  FORMAT(5X,2(I5,5X))
    30  FORMAT(5X,4( D20.6,5X))
```

```
 40  FORMAT(5X,'        ')
 50  FORMAT(3F10.0,2I5)
     READ(25,10)NUMEL,NTOT
     DO150 I=1,NTOT
     D(I)=0.0
     DO150 J=1,NTOT
150  GK(I,J)=0.0
     WRITE(26,40)
     WRITE(26,20)NUMEL,NTOT
     WRITE(26,40)
     DO350,I=1,NUMEL
     READ(25,50)T,AL,DIST,N1,N2
     WRITE (26,60)T,AL,DIST,N1,N2
 60  FORMAT(5X,3(D20.6,5X),2(I5,5X))
     GK(N1,N1)=GK(N1,N1)+T/AL
     GK(N2,N2)=GK(N1,N2)-T/AL
     GK(N2,N1)=GK(N2,N1)-T/AL
     GK(N2,N2)=GK (N2,N2)+T/AL
     D(N1)=D(N1)-T*DIST
     D(N2)=D(N2)+T*DIST
350  CONTINUE
*    WRITE(26,40)
     DO805 1I=1,NTOT
805  WRITE(26,30)(GK(I,J),J=1,NTOT)
     WRITE(26,40)
     WRITE(26,30)(D(I),I=1,NTOT)
     WRITE(26,40)
     READ(25,35)NUMR,(NDOF(I),I=1,NUMR)
 35  FORMAT(11I5)
     DO105 II=1,NUMR
     L=NDOF(I)
     D(L)=0.0
     DO106 J=1,NTOT
     GK(L,I)=0.0
106  GK(J,-1)=0.0
     GK(L,-1)=1.0
105  CONTINUE
     CALL INVERS(GK,C,NTOT)
     DO400 I=1,NTOT
     R(I)=0.0
     DO400 J=1,NTOT
400  R(I)=R(I)+GK(I,J)*D (J)
     WRITE(26,40)
     WRITE(26,30)(R(I),I=1,NTOT)
     WRITE (26,40)
     STOP
     END
       SUBROUTINE INVERS(A,U,N)
     IMPLICIT REAL*8(A-H,D-Z)
     DIMENSION A(20,20),U(20,20)
     DO1 I=1,N
```

```
        DO1J=1,N
        U(I,J)=0.0
        IF(I,EQ.J )U(I,J)=1.0
      1 CONTINUE
        EPS=0.0000001
        DO15 I=1,N
        K=I
        IF(I-N)21,7,21
     21 IF(A(I,I)-EPS)5,6,7
      5 IF(-A(I,I)-EPS)6,6,7
      6 K=K+1
        DO23 J=1,N
        U(I,J)=U(I,J)+U(K,J)
     23 A(I,J)=A(I,J)+A(K,J)
        GOTO21
      7 DIV=A(I,I)
        DO 9 I=1,N
        U(I,J)=U(I,J)/DIV
      9 A(I,J)=A(I,J)/DIV
        DO15 M=1,N
        DELT=A(M,I)
        IF(DABS(DELT)-EPS)15,15,16
     16 IF(M,I)10,15,10
     10 DO11 J=1,N
        U(M,J)=U(M,J)-U(I,J)*DELT
     11 A(M,J)=A(M,J)-A( I,J)*DELT
     15 CONTINUE
        DO34 I=1,N
        DO34 J=1,N
     34 A(I,J)=U(I,J)
        RETURN
        END
```

9. The thin-walled cross section contains 24 elements and 18 nodes (only the right half of the selection need be considered). Thus NUMEL = 24 and NTOT = 18. Two nodes, viz. 17 and 18, which are on the line of symmetry have zero warping. The complete input data is shown in the following:

Card No	Format	Data entered on card
1	2 I5	24, 18
2	3F10.0, 2 I5	0.035, 3.33, 21.82, 1, 2
3	3F10.0, 2 I5	0.019, 4.8, 14.5, 2, 4
4	3F10.0, 2 I5	0.012, 4.8, 11.17, 1, 3
5	3F10.0, 2 I5	0.020, 3.33, 17.02, 3, 4
6	3F10.0, 2 I5	0.012, 4.8, 11.17, 3, 5
7	3F10.0, 2 I5	0.019, 4.8, 14.5, 4, 6
8	3F10.0, 2 I5	0.020, 3.33, 12.22, 5, 6

(continued)

(continued)

Card No	Format	Data entered on card
9	3F10.0, 2 I5	0.012, 4.8, 11.17, 5, 9
10	3F10.0, 2 I5	0.019, 4.8, 14.5, 6, 7
11	3F10.0, 2 I5	0.016, 3.33, 7.42, 9,7
12	3F10.0, 2 I5	0.019, 1.89, 14.5, 7, 8
13	3F10.0, 2 I5	0.021, 3.33, 5.53, 10, 8
14	3F10.0, 2 I5	0.016, 1.89, 11.17, 9, 10
15	3F10.0, 2 I5	0.016, 2.58, 7.42, 11, 9
16	3F10.0, 2 I5	0.021, 2.58, 5.53, 12, 11
17	3F10.0, 2 I5	0.016, 1.89, 8.59, 11, 12
18	3F10.0, 2 I5	0.016, 2.58, 7.42, 13, 11
19	3F10.0, 2 I5	0.021, 2.58, 5.53, 14, 12
20	3F10.0, 2 I5	0.016, 1.89, 6.01, 13, 14
21	3F10.0, 2 I5	0.016, 2.58, 7.42, 15, 13
22	3F10.0, 2 I5	0.021, 2.58, 5.53, 16, 14
23	3F10.0, 2 I5	0.016, 1.89, 3.43, 15, 16
24	3F10.0, 2 I5	0.016, 3.43, 7.42, 17, 15
25	3F10.0, 2 I5	0.021, 3.43, 5.53, 18, 16
26	3I5	2, 17, 18

The normalized unit warping at the various nodes given by the program are

$$\omega_{n1} = -168.51\,\text{m}^2 \qquad \omega_{n9} = 58.31$$

$$\omega_{n2} = -90.01 \qquad \omega_{n10} = 80.49$$

$$\omega_{n3} = -87.42 \qquad \omega_{n11} = 45.16$$

$$\omega_{n4} = -31.45 \qquad \omega_{n12} = 61.66$$

$$\omega_{n5} = -14.61 \qquad \omega_{n13} = 31.67$$

$$\omega_{n6} = 26.03 \qquad \omega_{n14} = 43.09$$

$$\omega_{n7} = 83.44 \qquad \omega_{n15} = 18.08$$

$$\omega_{n8} = 106.24 \qquad \omega_{n16} = 24.58$$

Using this normalized unit warping distribution, the value of I_Ω can be computed for this multicellular section. If the section is made of steel, we have $EI_\Omega = 2,450,000\,\mathrm{G\,Nm}^4$. It may be noted that this is the section we used in some of the example problems of this and the previous chapter.

Reference

1. Hughes, T.J.R.: The Finite Element Method. Prentice Hall Inc., New Jersey (1987)

Chapter 7
Stability Analysis of Thin-Walled Structures in Torsion

7.1 Introduction

Linear instability analysis of thin-walled structures which involves torsion is considered in this Chapter. The goal in a linear instability analysis is the bifurcation buckling load, i.e., the load at which two equilibrium configurations are possible, viz. the prebuckled configuration and a proximate configuration. Equilibrium equations are thus used on the proximate configuration and result in a set of homogeneous equations defining the eigenvalue problem. The eigenvalues are the critical loads, and the corresponding eigenvectors are the proximate equilibrium mode shapes. Obviously, the lowest of these loads is the buckling load of the structure [1].

An initially straight column such as the one shown in Fig. 7.1 has the buckling load P defined by the differential equation [2].

$$EI\frac{d^4u}{dz^4} + P\frac{d^2u}{dz^2} = 0 \tag{7.1}$$

where u is the lateral deflection. Equation (7.1) can be solved for various boundary conditions. For a column with simply supported (hinged) end conditions, we can get critical load

$$P_{cr} = \frac{\pi^2 EI_y}{L^2} \tag{7.2}$$

The critical load for a possible buckling in the y-direction is

$$P_{cr} = \frac{\pi^2 EI_x}{L^2} \tag{7.3}$$

The lesser of the two loads given by Eqs. (7.2) and (7.3) will control.

In the foregoing, the buckling load of the column is computed on the assumption that the proximate equilibrium configuration is a bending configuration about one of

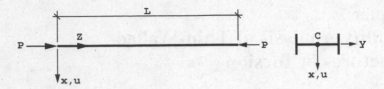

Fig. 7.1 Euler column

the principal axes. It is a buckling phenomenon involving only flexure of the axially loaded column and can thus be termed as flexural buckling.

Buckling of axially loaded columns can involve torsion also. Under axial load, a thin-walled initially straight bar can buckle by bending and twisting. This phenomenon is called flexural–torsional buckling. In short columns with doubly symmetric sections where the centroid and shear center coincide, the buckling under axial load can involve only twisting. This is called torsional buckling. In general with monosymmetric and unsymmetric sections, the buckling is the result of an interaction between flexure and torsion; i.e., it is of the flexural–torsional type.

Yet another form of buckling involving torsion occurs in beams and is called lateral buckling. This is in fact a flexural–torsional buckling. When the beam is loaded in the stiff principal plane, a proximate equilibrium shape can exist at a critical lateral load which consists of bending in the flexible plane accompanied by twisting.

In the following, solutions techniques for buckling problems involving torsion, described in the foregoing, are described. For simple boundary conditions and prismatic bars, closed-form solutions can be obtained using differential equations of the buckling problem. For complex cases, finite elements with elastic stiffness matrix described in the earlier chapter must be used together with a geometric stiffness matrix [3].

7.2   Flexural–Torsional Buckling

The differential equations of flexural–torsional buckling are derived in the following. From Eq. (7.1), we learn that

$$EI\frac{\mathrm{d}^4u}{\mathrm{d}z^4} = -P\frac{\mathrm{d}^2u}{\mathrm{d}z^2} \tag{7.4}$$

The foregoing shows that the effect of the axial load can be considered as the effect of a pseudo-lateral load of magnitude $P\frac{\mathrm{d}^2u}{\mathrm{d}z^2}$. This observation can be used to obtain the differential equations of flexural–torsional buckling. The cross section of a thin-walled member is shown in Fig. 7.2. The origin of coordinates is chosen at the centroid C. The shear center S is located at (x_0, y_0). The displacement of the section are translations ξ and η and a rotation ϕ about the shear center. After these

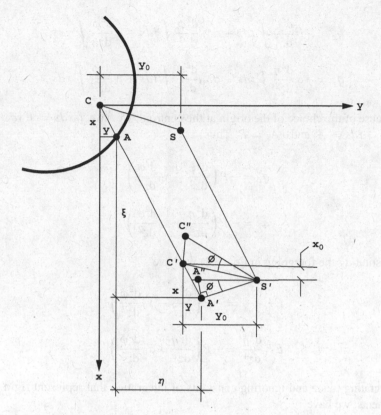

Fig. 7.2 Arbitrary cross section of the column

translations and rotation, C is located at C'' and S at S'. A general point A on the cross section is located at A''. An element at A has thus the following displacements.

$$u = \xi - (y_0 - y)\phi \tag{7.5}$$

$$v = \eta + (x_0 - x)\phi \tag{7.6}$$

The pseudo-lateral loads per unit length acting in the x- and y-directions are

$$q_x = \int -(\sigma t ds)\frac{d^2}{dz^2}[\xi - (y_0 - y)\phi]$$

$$q_y = \int -(\sigma t ds)\frac{d^2}{dz^2}[\eta + (x_0 - x)\phi]$$

where the integrals are over the contour of the cross section. Performing the integration, we have

$$q_x = -\sigma \frac{\mathrm{d}^2\xi}{\mathrm{d}z^2} \int t\,ds + \sigma y_0 \frac{\mathrm{d}^2\phi}{\mathrm{d}z^2} \int t\,ds - \sigma \frac{\mathrm{d}^2\phi}{\mathrm{d}z^2} \int y\,ds$$

$$q_y = -\sigma \frac{\mathrm{d}^2\eta}{\mathrm{d}z^2} \int t\,ds - \sigma x_0 \frac{\mathrm{d}^2\phi}{\mathrm{d}z^2} \int t\,ds + \sigma \frac{\mathrm{d}^2\phi}{\mathrm{d}z^2} \int x\,ds$$

Because of the choice of the origin at the centroid, we have $\int x\,ds = \int y\,ds = 0$. Further, $\int t\,ds = A_s$ and $\sigma A_s = P$. Thus,

$$q_x = -P\left(\frac{\mathrm{d}^2\xi}{\mathrm{d}z^2} - y_0\frac{\mathrm{d}^2\phi}{\mathrm{d}z^2}\right) \tag{7.7}$$

$$q_y = -P\left(\frac{\mathrm{d}^2\eta}{\mathrm{d}z^2} + x_0\frac{\mathrm{d}^2\phi}{\mathrm{d}z^2}\right) \tag{7.8}$$

Substituting the foregoing in Eq. (7.4) we have

$$EI_y\frac{\mathrm{d}^4\xi}{\mathrm{d}z^4} = -P\left(\frac{\mathrm{d}^2\xi}{\mathrm{d}z^2} - y_0\frac{\mathrm{d}^2\phi}{\mathrm{d}z^2}\right)$$

$$EI_x\frac{\mathrm{d}^4\eta}{\mathrm{d}z^4} = -P\left(\frac{\mathrm{d}^2\eta}{\mathrm{d}z^2} + x_0\frac{\mathrm{d}^2\phi}{\mathrm{d}z^2}\right)$$

Integrating twice and ignoring constants of integration that represent rigid body movements, we have

$$EI_y\frac{\mathrm{d}^2\xi}{\mathrm{d}z^2} + P\xi = Py_0\phi \tag{7.9}$$

$$EI_x\frac{\mathrm{d}^2\eta}{\mathrm{d}z^2} + P\eta = -Px_0\phi \tag{7.10}$$

We also have, about S',

$$m_z = \int -(\sigma t\,ds)(y_0 - y)\frac{\mathrm{d}^2}{\mathrm{d}z^2}[\xi - (y_0 - y)\phi]$$
$$+ \int (\sigma t\,ds)(x_0 - x)\frac{\mathrm{d}^2}{\mathrm{d}z^2}[\eta + (x_0 - x)\phi]$$

Integrating and using the known relations, we have

$$m_z = -Py_0\frac{\mathrm{d}^2\xi}{\mathrm{d}z^2} + Py_0^2\frac{\mathrm{d}^2\phi}{\mathrm{d}z^2} + \frac{P}{A_s}I_x\frac{\mathrm{d}^2\phi}{\mathrm{d}z^2}$$
$$+ Px_0\frac{\mathrm{d}^2\eta}{\mathrm{d}z^2} + Px_0^2\frac{\mathrm{d}^2\phi}{\mathrm{d}z^2} + \frac{P}{A_s}I_y\frac{\mathrm{d}^2\phi}{\mathrm{d}z^2}$$

$$= P\left[x_0\frac{d^2\eta}{dz^2} - y_0\frac{d^2\xi}{dz^2}\right]$$

$$+ \frac{P}{A_s}\left[A_s y_0^2 + A_s x_0^2 + I_x + I_y\right]\frac{d^2\phi}{dz^2}.$$

Introducing I_0, the polar moment of inertia about the shear center, defined as

$$I_0 = I_x + I_y + A_s\left(x_0^2 + y_0^2\right)$$

We have

$$m_z = P\left[x_0\frac{d^2\eta}{dz^2} - y_0\frac{d^2\xi}{dz^2}\right] + \frac{P I_0}{A_s}\frac{d^2\phi}{dz^2}$$

Using the foregoing in the differential equation for mixed torsion derived in Chap. 4, we have

$$EI_\Omega\frac{d^4\phi}{dz^4} - GJ\frac{d^2\phi}{dz^2} = -P\left[x_0\frac{d^2\eta}{dz^2} - y_0\frac{d^2\xi}{dz^2}\right] - \frac{P I_0}{A_s}\frac{d^2\phi}{dz^2}$$

Or

$$EI_\Omega\frac{d^4\phi}{dz^4} - \left(GJ - \frac{P I_0}{A_s}\right)\frac{d^2\phi}{dz^2} + Px_0\frac{d^2\eta}{dz^2} - Py_0\frac{d^2\xi}{dz^2} = 0 \qquad (7.11)$$

Equations (7.9), (7.10) and (7.11) define the differential equations of flexural–torsional buckling under an axial load P acting at the centroid of the thin-walled section. These equations show that they are coupled and hence the flexural buckling in the stiff principal plane, flexural buckling in the flexible principal plane, and the torsional buckling are all coupled when the shear center and the center of gravity do not coincide. There is thus an interaction between flexural and torsional buckling, and the actual buckling load may be less than any of these values.

When the shear center and the center of gravity coincide (as in the case of a doubly symmetric section), we have $x_0 = y_0 = 0$ and the equations are decoupled.

$$EI_y\frac{d^2\xi}{dz^2} + P\xi = 0 \qquad (7.12)$$

$$EI_x\frac{d^2\eta}{dz^2} + P\eta = 0 \qquad (7.13)$$

$$EI_\Omega\frac{d^4\phi}{dz^4} - \left(GJ - \frac{P I_0}{A_s}\right)\frac{d^2\phi}{dz^2} = 0 \qquad (7.14)$$

Equation (7.12) represents buckling in the x–z principal plane. The critical load is

$$P_y = \frac{\pi^2 E I_y}{L^2} \tag{7.15}$$

Equation (7.13) represents buckling in the y–z principal plane. The critical load is

$$P_x = \frac{\pi^2 E I_x}{L^2} \tag{7.16}$$

Equation (7.14) describes pure torsional buckling and the critical axial load is given by

$$P_\phi = \frac{A_s}{I_0} \left(GJ + \frac{\pi^2 E I_\Omega}{L^2} \right) \tag{7.17}$$

For monosymmetric beams such as, for example, beams symmetrical about the x-axis for which $y_0 = 0$, the differential equations of flexural–torsional buckling are

$$E I_y \frac{d^2 \xi}{dz^2} + P\xi = 0 \tag{7.18}$$

$$E I_x \frac{d^2 \eta}{dz^2} + P\eta = -P x_0 \phi \tag{7.19}$$

$$E I_\Omega \frac{d^4 \phi}{dz^4} - \left(GJ - \frac{P I_0}{A_s} \right) \frac{d^2 \phi}{dz^2} + P x_0 \frac{d^2 \eta}{dz^2} = 0 \tag{7.20}$$

Equations (7.19) and (7.20) are coupled. Their solution can be assumed in the form.

$$\eta = A_2 \sin \frac{\pi z}{L}$$

$$\phi = A_3 \sin \frac{\pi z}{L}$$

The critical load is then given by

$$(P - P_x) A_2 + P x_0 A_3 = 0$$

$$P x_0 A_2 + \frac{I_0}{A_s} (P - P_\phi) = 0$$

The critical value of P is given by

$$\begin{vmatrix} P - P_x & Px_0 \\ Px_0 & \frac{I_0}{A_s}\left(P - P_\phi\right) \end{vmatrix} = 0$$

Which results in a quadratic equation in P whose solution can be written as

$$P = \frac{\frac{I_0}{A_s}\left(P_x + P_\phi\right) \pm \sqrt{\left\{\frac{I_0^2}{A_s^2}\left(P_x + P_\phi\right)^2 - 4P_x P_\phi \frac{I_0}{A_s}\left(\frac{I_0}{A_s} - x_0^2\right)\right\}}}{2\left(\frac{I_0}{A_s} - x_0^2\right)} \tag{7.21}$$

For the most general case of an unsymmetric section, we can assume the solution as

$$\xi = A_1 \sin\frac{\pi z}{L}$$

$$\eta = A_2 \sin\frac{\pi z}{L}$$

$$\phi = A_3 \sin\frac{\pi z}{L}$$

Using the foregoing in Eqs. (7.9), (7.10), and (7.11) we have the following homogenous equations defining the eigenvalue problem.

$$\left(P - P_y\right)A_1 - Py_0 A_3 = 0$$

$$\left(P - P_x\right)A_2 + Px_0 A_3 = 0$$

$$-Py_0 A_1 + Px_0 A_2 + \left(P - P_\phi\right)A_3 = 0$$

The solution is given by the following

$$\begin{vmatrix} P - P_y & 0 & -Py_0 \\ 0 & P - P_x & Px_0 \\ -Py_0 & Px_0 & P - P_\phi \end{vmatrix} = 0$$

The foregoing gives a cubic equation for P

$$\left(\frac{I_x + I_y}{I_0}\right)P^3 + \left\{\frac{A_s}{I_0}\left(P_x y_0^2 + P_y x_0^2\right) - \left(P_x + P_y + P_\phi\right)\right\}P^2$$
$$+ \left(P_x P_y + P_x P_\phi + P_y P_\phi\right)P - P_x P_y P_\phi = 0 \tag{7.22}$$

Fig. 7.3 Example 7.1

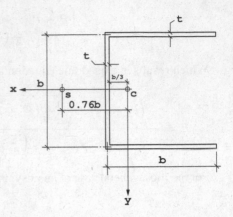

The critical load resulting from the interaction of flexural and torsional buckling will be less than any of the three pure values, and hence, the interaction must be considered in the design.

Example 7.1 A channel section is used as a column 3 m long with hinged ends. The three sides of the channel are of equal length and thickness as shown in Fig. 7.3. What are the limiting dimensions of the cross section? Use $E = 2.5G$ and $E = 1000\sigma_y$. Assume free warping so that St. Venant torsion only exists.

The various section properties that can be computed are

$$I_y = \frac{b^3 t}{3} \quad I_x = \frac{7b^3 t}{12} \quad J = bt^3$$

$$I_0 = \frac{b^3 t}{3} + \frac{7b^3 t}{12} + 3bt(0.76b)^2 = 2.65b^3 t$$

The Euler load is given by

$$P_E = \frac{\pi^2 E \left(\frac{b^3 t}{3}\right)}{L^2}$$

Equating this to yield load

$$\frac{\pi^2 E b^3 t}{3L^2} = \frac{3bt E}{1000}$$

which gives

$$\frac{L}{b} = 33.12$$

The torsional buckling load is given by

$$P_T = \frac{A_s}{I_0}GJ = \frac{3bt}{2.65b^3t}\frac{E}{2.5}bt^3 = 0.4528E\left(\frac{t^3}{b}\right)$$

Equating this to the yield load

$$0.4528E\left(\frac{t^3}{b}\right) = \frac{3btE}{1000}$$

which gives

$$\frac{b}{t} = 12.29$$

A possible design of the section is thus

$$b = \frac{3}{33.12} = 0.091\,\text{m}$$

And

$$t = \frac{0.091}{12.29} = 0.0074\,\text{m}$$

Hence, $b = 91$ mm and $t = 7.4$ mm. With this design torsional buckling (with free warping), Euler buckling and yielding will all occur simultaneously. The design will not be satisfactory when warping is restrained.

Example 7.2 A 4.5 m long simply supported steel column has the thin-walled section as shown in Fig. 7.4.

The sectional properties are $A_s = 2420\,\text{mm}^2$, $I_y = 3{,}675{,}320\,\text{mm}^4$, and $I_x = 953{,}170\,\text{mm}^4$. Show that the column buckles in bending rather than in combined bending and torsion.

$$E = 2 \times 10^5\,\text{N/mm}^2 \text{ and } G = 0.79 \times 10^5\,\text{N/mm}^2.$$

The following sectional properties can be computed

$$J = \frac{2}{3}(100)(12.5)^3 = 130{,}208\,\text{mm}^4$$

$$I_\Omega = \frac{(12.5)^3(100)^3}{18} = 108{,}506{,}944\,\text{mm}^6$$

It may be noted that the contour warping of the section is zero but the thickness warping is considered in the calculation of I_Ω.

Fig. 7.4 Example 7.2

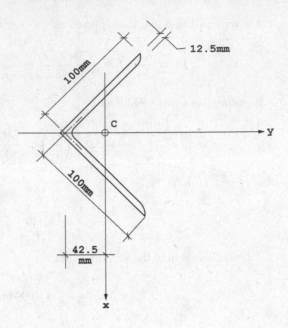

The Euler buckling load is given by

$$P_E = \frac{\pi^2(2 \times 10^5)(953,170)}{(4500)^2} = 92,913 \, \text{N}$$

The polar moment of inertia is

$$I_0 = 3,675,320 + 953,170 + 2420(42.5)^2 = 8,999,615 \, \text{mm}^4$$

We have

$$P_y = \frac{\pi^2(2 \times 10^5)(3,675,320)}{(4500)^2} = 358,262 \, \text{N}$$

$$\frac{I_0}{A_s} - y_o^2 = \frac{8,999,615}{2420} - (42.5)^2 = 1912.6$$

The pure torsional buckling load is

$$P_\phi = \frac{2420}{8,999,615}\left[0.79 \times 10^5 \times 130,208 + \frac{\pi^2(2 \times 10^5)(108,506,944)}{(4500)^2}\right]$$

$$= 2,768,870 \, \text{N}$$

(The warping contribution is found to be only 0.1% of St. Venant) The flexural–torsional buckling load is thus

$$P = \frac{(3718.8)(3,127,132) \pm \sqrt{1.3523779 \times 10^{20} - 4(358,262)(2,768,870)(3718.8)(1912.6)}}{2(1912.6)}$$

$$= \frac{1.1629178 \times 10^{10} \pm 1.0344836 \times 10^{10}}{2(1912.6)}$$

$$= 335,758\,\text{N}.$$

The column thus fails only in pure flexure about the minor axis.

Example 7.3 In Fig. 7.4 if the wall thickness is 6.25 mm and the length of column is 1.5 m, show that it would buckle in combined bending and torsion. The section properties are $y_0 = -39\,\text{mm}$, $A_s = 1252\,\text{mm}^2$, $I_y = 2,018,722\,\text{mm}^4$, and $I_x = 510,300\,\text{mm}^4$. We have

$$P_E = \frac{\pi^2 (2 \times 10^5)(510,300)}{(1500)^2} = 447,685\,\text{N}$$

$$I_0 = 2,018,722 + 510,300 + 1252(39)^2 = 4,433,314\,\text{mm}^4$$

$$\frac{I_0}{A_s} - y_o^2 = \frac{4,433,314}{1252} - 39^2 = 2020$$

$$P_y = \frac{\pi^2 (2 \times 10^5)(2,018,722)}{(1500)^2} = 1,771,021\,\text{N}$$

$$P_\phi = \frac{1252}{4,433,314}\left[0.79 \times 10^5 \times (16,276) + \frac{\pi^2 (2 \times 10^5)(13,563,368)}{(1500)^2} \right]$$

$$= 366,481\,\text{N}$$

(The warping contribution is 1% of St. Venant).
The flexural–torsional buckling load is

$$P = \frac{(3541)(2,137,502) \pm \sqrt{5.7288165 \times 10^{19} - 1.8570024 \times 10^{19}}}{2(2020)}$$

$$= 333,293\,\text{N} < P_E$$

Example 7.4 A channel column with the cross section shown in Fig. 7.5 is 3 m long and has the following properties.

$$A_s = 2065\,\text{mm}^2, \ J = 17,770\,\text{mm}^4, \ I_\Omega = 16,031,591,000\,\text{mm}^6,$$

Fig. 7.5 Example 7.4

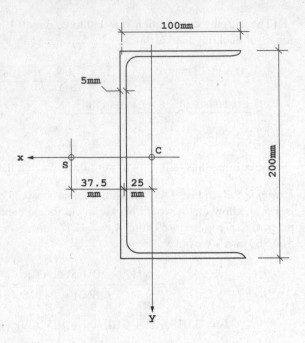

$$I_x = 14{,}193{,}490 \text{ mm}^4 \text{ and } I_y = 2{,}222{,}680 \text{ mm}^4.$$

Determine the buckling load of the column.
We have

$$P_E = \frac{\pi^2 (2 \times 10^5)(2{,}222{,}680)}{(3000)^2} = 487{,}488 \text{ N}$$

$$P_x = \frac{\pi^2 (2 \times 10^5)(14{,}193{,}490)}{(3000)^2} = 3{,}112{,}981 \text{ N}$$

$$I_0 = 14{,}193{,}490 + 2{,}222{,}680 + 2065(62.5)^2 = 24{,}482{,}576 \text{ mm}^4$$

$$\frac{I_0}{A_s} - x_0^2 = 11{,}856 - 3906 = 7950$$

$$P_\phi = \frac{1}{11{,}856} \left[0.79(10^5)(17{,}770) + \frac{\pi^2 (2 \times 10^5)(16{,}031{,}591{,}000)}{(3000)^2} \right]$$

$$= 414{,}976 \text{ N}$$

In the foregoing, it can be seen that the warping contribution is significant. The flexural–torsional buckling load is

$$P = \frac{(11{,}856)(3{,}527{,}957) \pm \sqrt{17.495363(10^{20}) - 4.8704015(10^{20})}}{2(7950)}$$

$$= 395{,}965\,\text{N} < P_E$$

The column thus fails in flexural–torsional buckling, and the buckling load is 395,965 N.

7.2.1 Stiffeners for Torsional Buckling

The torsional buckling resistance of thin-walled section columns can be improved by providing stiffeners as shown in Fig. 7.6. The warping restraint stiffness, K, of the stiffener at the stiffened section can be given as

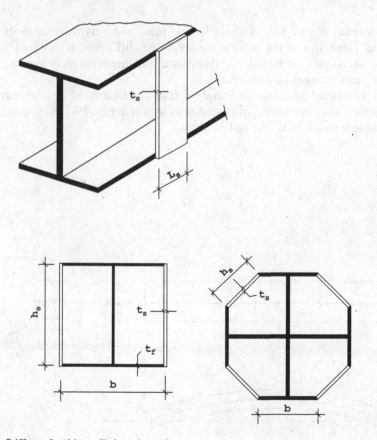

Fig. 7.6 Stiffness for thin-walled section column

$$K = \frac{12\left(\frac{b}{h_s}\right)^2 \left(\frac{1}{h_s}\right)\left(\frac{I_s}{I_f}\right)}{\left[1 + 12\left(\frac{E}{G}\right)\frac{I_s}{A_{st}h_s^2}\right]}$$ (7.23)

where A_{st} is the area of the stiffener equal to $L_s t_s$ and

$$I_s = \frac{1}{12} t_s L_s^3$$

$$I_f = \frac{1}{12} t_f b^3$$

7.3 Lateral Buckling of Thin-Walled Beams

When a beam is very stiff in one principal plane and very flexible in the other principal plane, then if the beam is loaded in the stiff plane, it can buckle in the flexible direction at a critical load. This lateral buckling consists of bending in the flexible plane accompanied by twisting.

The differential equations for lateral buckling can be derived by considering the beam with doubly symmetric cross section as shown in Fig. 7.7. The principal axes rotate to new positions x_1, y_1, and z_1. Thus,

$$EI_x \frac{d^2\eta}{dz^2} = M_{x1}$$

$$EI_y \frac{d^2\xi}{dz^2} = M_{y1}$$

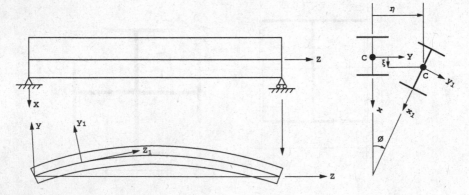

Fig. 7.7 Lateral breaking of doubly symmetric cross section

Fig. 7.8 Simply supported beam in pure bending

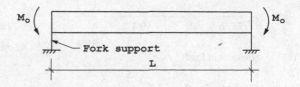

$$EI_\Omega \phi'''' - GJ\phi'' = m_{z1} \tag{7.24}$$

where the right-hand sides consist of bending and twisting moments resolved along the rotated principal axes.

As an example, the simply supported beam in pure bending M_0 shown in Fig. 7.8 is considered. We have

$$M_{x1} = M_0\phi; \quad M_{y1} = M_0 \text{ and } M_{z1} = -M_0\frac{d\eta}{dz}$$

Hence,

$$EI_x \frac{d^2\eta}{dz^2} = M_0\phi$$

$$EI_y \frac{d^2\xi}{dz^2} = M_0$$

$$EI_\Omega\phi'''' - GJ\phi'' = M_0\frac{d^2\eta}{dz^2}$$

The solution of these partially coupled equations gives the critical value of M_0 that causes lateral buckling.

$$M_0 = \frac{\pi}{L}\sqrt{EI_\Omega GJ\left(1 + \frac{\pi^2 EI_\Omega}{GJL^2}\right)} \tag{7.25}$$

For deep, narrow beams we have simply

$$M_0 = \frac{\pi}{L}\sqrt{EI_\Omega GJ} \tag{7.26}$$

Results for other standard cases are available in the literature. A few are summarized in the following

1. Simply supported beam with a central load P

$$P = \frac{\alpha}{L^2}\sqrt{EI_\Omega GJ} \tag{7.27}$$

where typical α values are

$L^2 GJ/EI_\Omega$	0.4	4	8	16	24	48	400
α (Centroidal load)	86.4	31.9	25.6	21.8	20.3	18.8	17.2
α (Topflange load)	51.5	–	–	–	–	–	15.8
α (bottomflange load)	147.0	–	–	–	–	–	18.7

2. Cantilever subjected to end load P

$$P = \frac{4.01}{L^2} \sqrt{EI_\Omega GJ} \tag{7.28}$$

7.4 Buckling Under Torsional Loadings

The earlier sections covered buckling under axial loading and lateral loading, i.e., columns and beams. Buckling of long members can also occur under pure torsional loading. At the critical torsional moment, the member can buckle into a space curve. The critical twisting moment is given by

$$M_{tc} = \frac{2\pi EI}{L} \tag{7.29}$$

When an axial compression exists with torsion, we have the buckling governed by

$$\left(\frac{M_t}{M_{tc}}\right)^2 + \frac{P}{P_E} = 1 \tag{7.30}$$

Buckling under torsion does not have much practical application. However, these calculations may be needed for long drill strings used in oil well drilling.

7.5 Finite Element Analysis of Buckling Problems

In the finite element analysis of buckling problems, two stiffness matrices are used. The elastic stiffness matrix is

$$[k] = \int [B]^T [D][B] \mathrm{d}v \tag{7.31}$$

which is explained in the previous chapter. The additional matrix is called the geometric stiffness matrix and is due to the interaction of prebuckling forces on the buckling configuration. A general expression for the geometric stiffness matrix is

$$[k_g] = \int [G]^{\mathrm{T}}[\sigma][G]\mathrm{d}v \tag{7.32}$$

where $[\sigma]$ is the matrix of prebuckling stresses and contains only the load and geometrical dimensions of the element. The matrix $[G]$ is the rotation–displacement matrix and is due to the nonlinear terms in the strain–displacement relationship. It may be noted that $[B]$ is the strain–displacement matrix and is due to the linear terms in the strain–displacement relationship.

The buckling phenomenon is governed by the relation

$$[[k] + [k_g]]\{\delta\} = \{0\} \tag{7.33}$$

where $\{\delta\}$ is the vector of buckling displacements. The system of homogenous equations represented by Eq. (7.33) will have a solution when

$$|[k] + [k_g]| = 0 \tag{7.34}$$

The foregoing equation gives the polynomial equation in terms of the critical load, the lowest root being the answer to the buckling problem. In practice, the polynomial equation is not solved but an iterative solution of Eq. (7.34) is used, viz. assuming a buckling load and evaluating the determinant and improving this process until the determinant is zero.

A thin-walled member finite element is shown in Fig. 7.9. The displacement vector at node 1 in the proximate equilibrium configuration is

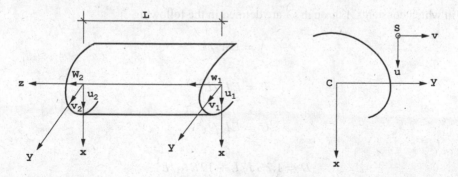

Fig. 7.9 Thin-walled beam finite element

$$\{\delta\}_1 = \begin{Bmatrix} w_1 \\ u_1 \\ v_1 \\ \theta_{z1} \\ \theta_{x1} \\ \theta_{y1} \\ \theta'_{z1} \end{Bmatrix} \tag{7.35}$$

Here w_1 is the axial displacement, u_1 and θ_{y1} represent bending in the xz principal plane, v_1 and θ_{x1} represent the bending in the yz principal plane and θ_{z1} and θ'_{z1} are the torsional rotation and warping, respectively. The displacement vector

$$\{\delta\} = \left\{ \frac{\{\delta\}_1}{\{\delta\}_2} \right\} \tag{7.36}$$

The stiffness matrix $[k]$ can be written as

$$[k] = \begin{bmatrix} A & 0 & 0 & 0 & 0 & 0 & 0 & -A & 0 & 0 & 0 & 0 & 0 & 0 \\ & 12B & 0 & 0 & 0 & 6LB & 0 & 0 & -12B & 0 & 0 & 0 & 6LB & 0 \\ & & 12C & 0 & -6LC & 0 & 0 & 0 & 0 & -12C & 0 & -6LC & 0 & 0 \\ & & & D & 0 & 0 & F & 0 & 0 & 0 & -D & 0 & 0 & F \\ & & & & 4L^2C & 0 & 0 & 0 & 0 & 6LC & 0 & 2L^2C & 0 & 0 \\ & & & & & 4L^2B & 0 & 0 & -6LB & 0 & 0 & 0 & 2L^2B & 0 \\ & & & & & & E & 0 & 0 & 0 & -F & 0 & 0 & G' \\ & & & & & & & A & 0 & 0 & 0 & 0 & 0 & 0 \\ & \text{sym} & & & & & & & 12B & 0 & 0 & 0 & -6LB & 0 \\ & & & & & & & & & 12C & 0 & 6LC & 0 & 0 \\ & & & & & & & & & & D & 0 & 0 & -F \\ & & & & & & & & & & & 4L^2C & 0 & 0 \\ & & & & & & & & & & & & 4L^2B & 0 \\ & & & & & & & & & & & & & E \end{bmatrix} \tag{7.37}$$

in which constants A through G' are defined in the following

$$A = EA_s/L$$

$$B = EI_y/L^3$$

$$C = EI_x/L^3$$

$$D = 1.2GJ/L + 12EI_\Omega/L^3$$

$$E = 2GJL/15 + 4EI_\Omega/L$$

$$F = GJ/10 + 6EI_\Omega/L^2$$

$$G' = -GJL/30 + 2EI_\Omega/L$$

The geometric stiffness matrix of the finite element can be written as

$$[k_g] = \begin{bmatrix}
0 & 0 & 0 & 0 & 0 & 0 & 0 & 0 & 0 & 0 & 0 & 0 & 0 \\
& a & 0 & V_1 & -e & b & V_4 & 0 & -a & 0 & V_7 & e & b & V_{10} \\
& & a & W_1 & -b & -e & W_4 & 0 & 0 & -a & W_7 & -b & e & W_{10} \\
& & & \theta_1 & -W_2 & V_2 & \theta_2 & 0 & -V_1 & -W_1 & \theta_3 & -W_3 & V_3 & \theta_4 \\
& & & & c & 0 & -W_5 & 0 & e & b & -W_8 & -d & f & -W_{11} \\
& & & & & c & V_5 & 0 & -b & e & V_8 & f & -d & V_{11} \\
& & & & & & \theta_5 & 0 & -V_4 & -W_4 & \theta_6 & -W_6 & V_6 & \theta_7 \\
& & & & & & & 0 & 0 & 0 & 0 & 0 & 0 & 0 \\
& & & & & & & & a & 0 & -V_7 & -e & -b & -V_{10} \\
& & & & & & & & & a & -W_7 & b & -e & -W_{10} \\
& & & & & & & & & & \theta_1 & -W_9 & V_9 & \theta_8 \\
& & & & & & & & & & & c & 0 & -W_{12} \\
& & & & & & & & & & & & c & V_{12} \\
& & & & & & & & & & & & & \theta_9
\end{bmatrix} \qquad (7.38)$$

In which the constants are defined in the following

$$a = \frac{1.2P_z}{L} \quad b = \frac{P_z}{10} \quad c = \frac{2P_zL}{15} \quad d = \frac{P_zL}{30} \quad e = \frac{M_{z1}}{L} \quad f = \frac{M_{z1}}{2}$$

$$A_1 = P_zy_0 + M_{x1} \quad B_1 = -P_zx_0M_{y1} \quad C_1 = \frac{P_zI_0}{A_s} - \beta_1M_{x1} + \beta_2M_{y1}$$

$$D_1 = \beta_1F_{y1} + \beta_2F_{x1} \quad \beta_1 = \frac{I}{I_x}\int y(x^2 + y^2)dA - 2y_0$$

$$\beta_2 = \frac{I}{I_y}\int x(x^2 + y^2)dA - 2x_0$$

$$V_1 = 1.2A_1 + \frac{F_{y1}}{10} \quad V_2 = \frac{A_1}{10} + \frac{2LF_{y1}}{10} \quad V_3 = \frac{A_1}{10} - \frac{LF_{y1}}{10} \quad V_4 = \frac{A_1}{10}$$

$$V_5 = \frac{2LA_1}{15} + \frac{L^2F_{y1}}{30} \quad V_6 = -\frac{LA_1}{30} - \frac{L^2F_{y1}}{30} \quad V_7 = -1.2\frac{A_1}{L} - \frac{11}{10}F_{y1}$$

$$V_8 = -\frac{A_1}{10} - 0.2LF_{y1}$$

$$V_9 = -\frac{A_1}{10} + 0.1LF_{y1} \quad V_{10} = \frac{A_1}{10} + 0.1LF_{y1} \quad V_{11} = -\frac{LA_1}{30}$$

$$V_{12} = \frac{2LA_1}{15} + 0.1L^2F_{y1}$$

W_1 Through W_{12} are obtained by replacing A_1 by B_1 and F_{y1} by $-F_{x1}$ in the expressions for V_1 through V_{12}.

$$\theta_1 = 1.2\frac{C_1}{L} - 0.6D_1 \quad \theta_2 = 0.1C_1 - 0.1LD_1 \quad \theta_3 = -1.2\frac{C_1}{L} + 0.6D_1$$

$$\theta_4 = 0.1C_1 \quad \theta_5 = \frac{2LC_1}{15} - \frac{L^2D_1}{30} \quad \theta_6 = -\frac{C_1}{10} + 0.1LD_1$$

$$\theta_7 = -\frac{LC_1}{30} + \frac{L^2D_1}{60} \quad \theta_8 = -0.1C_1 \quad \theta_9 = \frac{2LC_1}{15} - 0.1L^2D_1$$

The use of the foregoing stiffness matrices in the buckling analysis of thin-walled members is illustrated in the following two examples.

Example 7.5 Determine the torsional buckling load of the thin-walled member with an angle section shown in Fig. 7.10. Use two finite elements to model the member.

Fig. 7.10 Example 7.5

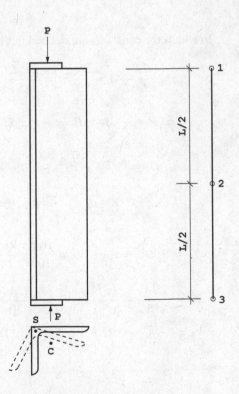

The only degrees of freedom which are of interest in this problem are $\theta_{z1}, \theta'_{z1}$ and $\theta_{z2}, \theta'_{z2}$. Due to symmetry only half of the structure is considered. The boundary conditions are $\theta_{z1} = \theta'_{z2} = 0$. Thus, the total stiffness is

$$[K] = \begin{bmatrix} \left(\frac{8EI_\Omega}{L} + \frac{GJL}{15} - \frac{PLI_0}{15A_s}\right) & \left(-\frac{GJ}{10} - \frac{24EI_\Omega}{L^2} + \frac{PI_0}{10A_s}\right) \\ \text{SYM} & \left(\frac{2.4GJ}{L} + \frac{96EI_\Omega}{L^3} - 2.4\frac{PI_0}{LA_s}\right) \end{bmatrix}$$

Hence, the critical load P_ϕ is given by

$$|[K]| = 0$$

$$\left(\frac{8EI_\Omega}{L} + \frac{GJL}{15} - \frac{P_\phi LI_0}{15A_s}\right)\left(\frac{2.4GJ}{L} + \frac{96EI_\Omega}{L^3} - 2.4\frac{P_\phi I_0}{LA_s}\right)$$

$$= \left(\frac{P_\phi I_0}{10A_s} - \frac{GJ}{10} - \frac{24EI_\Omega}{L^2}\right)^2$$

which gives, after the necessary algebra, a quadratic equation in P_ϕ (GJ terms are neglected)

$$P_\phi^2 - \frac{416}{3}\frac{EI_\Omega A_s}{I_0 L^2} P_\phi + 1280\left(\frac{EI_\Omega A_s}{I_0 L^2}\right)^2 = 0$$

The solution of the quadratic equation gives

$$P_\phi = 9.945\frac{EI_\Omega A_s}{I_0 L^2}$$

which can be compared with the exact value given by Eq. (7.17), viz.

$$P_\phi = 9.87\frac{EI_\Omega A_s}{I_0 L^2}.$$

Example 7.6 Determine the critical bending moment M_0 that would cause lateral buckling in the beam shown in Fig. 7.8. Use two finite elements.

Because of symmetry, only one half of the structure is considered. The degrees of freedom relevant to the problem are $V_1, \theta_{z1}, \theta_{x1}$, and θ'_{z1} and $V_2, \theta_{z2}, \theta_{x2}$ and θ'_{z2}. The boundary conditions are $V_1 = \theta_{z1} = \theta_{x2} = \theta'_{z2} = 0$. Thus, the total stiffness is (neglecting GJ terms).

$$[K] = \begin{bmatrix} \frac{8EI_x}{L} & -\frac{M_0 L}{15} & \frac{24EI_x}{L^2} & \frac{M_0}{10} \\ -\frac{M_0 L}{15} & \frac{8EI_\Omega}{L} & -\frac{M_0}{10} & \frac{24EI_\Omega}{L^2} \\ \frac{24EI_x}{L^2} & -\frac{M_0}{10} & \frac{96EI_x}{L^3} & \frac{2.4M_0}{L} \\ \frac{M_0}{10} & -\frac{24EI_\Omega}{L^2} & \frac{2.4M_0}{L} & \frac{96EI_\Omega}{L^3} \end{bmatrix}$$

Equating the determinant to zero, we get

$$
\frac{8EI_x}{L}
\begin{vmatrix}
\frac{8EI_\Omega}{L} & -\frac{M_0}{10} & -\frac{24EI_\Omega}{L^2} \\
-\frac{M_0}{10} & \frac{96EI_x}{L^3} & \frac{2.4M_0}{L} \\
-\frac{24EI_\Omega}{L^2} & \frac{2.4M_0}{L} & \frac{96EI_\Omega}{L^3}
\end{vmatrix}
+ \frac{M_0L}{15}
\begin{vmatrix}
-\frac{M_0L}{15} & -\frac{M_0}{10} & -\frac{24EI_\Omega}{L^2} \\
\frac{24EI_x}{L^2} & \frac{96EI_x}{L^3} & \frac{2.4M_0}{L} \\
\frac{M_0}{10} & \frac{2.4M_0}{L} & \frac{96EI_\Omega}{L^3}
\end{vmatrix}
$$

$$
+ \frac{24EI_x}{L^2}
\begin{vmatrix}
-\frac{M_0L}{15} & \frac{8EI_\Omega}{L} & -\frac{24EI_\Omega}{L^2} \\
\frac{24EI_x}{L^2} & -\frac{M_0}{10} & \frac{2.4M_0}{L} \\
\frac{M_0}{10} & -\frac{24EI_\Omega}{L^2} & \frac{96EI_\Omega}{L^3}
\end{vmatrix}
- \frac{M_0}{10}
\begin{vmatrix}
-\frac{M_0L}{15} & \frac{8EI_\Omega}{L} & -\frac{M_0}{10} \\
\frac{24EI_x}{L^2} & -\frac{M_0}{10} & \frac{96EI_x}{L^3} \\
\frac{M_0}{10} & -\frac{24EI_\Omega}{L^2} & \frac{2.4M_0}{L}
\end{vmatrix}
$$

which results in the quadratic equation

$$
0.0225 M_0^4 - 375.04\left(\frac{E^2 I_x I_\Omega}{L^4}\right) M_0^2 + 36864\left(\frac{E^2 I_x I_\Omega}{L^4}\right)^2 = 0
$$

Hence,

$$
M_0^2 = \left(\frac{375.04 - 370.59}{0.045}\right)\frac{E^2 I_x I_\Omega}{L^4}
$$

$$
= 98.88889 \frac{E^2 I_x I_\Omega}{L^4}
$$

Thus,

$$
M_0 = 9.944 \frac{E}{L^2}\sqrt{I_x I_\Omega}
$$

This can be compared with the exact value

$$
M_0 = 9.87 \frac{E}{L^2}\sqrt{I_x I_\Omega}.
$$

References

1. McGuire, W., Gallagher, R.H., Ziemian, R.D.: Matrix Structural Analysis. Wiley, New York (2000)
2. Simitses, G.J., Hodges, D.H.: Fundamentals of Structural Stability. Elsevier Inc., Amsterdam (2006)
3. Rajagopalan, K.: Finite Element Buckling Analysis of Stiffened Cylindrical Shells. CRC Press, Boca Raton (1983)

Chapter 8
Plastic Torsion of Thin-Walled Structures

8.1 Introduction

Torsion of thin-walled structures when parts of the structure yield in shear can be tackled by the finite element method. The stiffness of the finite element due to material nonlinearities must be defined suitably. However, in many design situations, it is only necessary to know the collapse torsional load. The collapse load can be predicted more easily, as in the case of bending, using rigid-plastic assumptions.

8.2 Ultimate Torque of a Thin Rectangle

The ultimate torque in St. Venant torsion can be predicted by the yield stress distribution shown in Fig. 8.1. The 'turning-around' behavior is consistent with the sand-heap analogy proposed for plastic torsion. The yield shear stress distribution in each orthogonal direction contributes to one-half of the total ultimate torque as in elastic torsion. The ultimate torque is thus

$$T_u = \tau_y \left(\frac{bt}{2}\right)\frac{t}{2} + \tau_y \left[\frac{1}{2}t\frac{t}{2}\right]b$$
$$= \tau_y \left(\frac{bt^2}{4}\right) + \tau_y \left(\frac{bt^2}{4}\right)$$
$$= \tau_y \frac{bt^2}{2} \tag{8.1}$$

© The Author(s), under exclusive license to Springer Nature Singapore Pte Ltd. 2022
K. Rajagopalan, *Torsion of Thin Walled Structures*,
https://doi.org/10.1007/978-981-16-7458-7_8

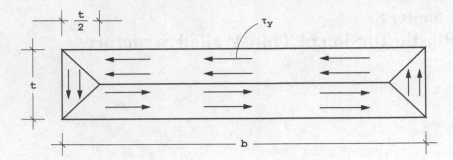

Fig. 8.1 Yield stress distribution in thin rectangle

8.3 Ultimate St. Venant Torque of Closed Sections

In closed or cellular sections, two types of yield shear stress distributions are possible. At the ultimate torque, a few (especially the thinner) walls will have a distribution shown in Fig. 8.2a while in the remaining walls, the distribution will be as shown in Fig. 8.2b. The former distribution can be called a 'shear band.' The walls with shear bands must be located by trial and error. The shear band shows that the plastic neutral axis is located outside the section in those contours with the shear bands [1].

As an example the thin-walled circular cross section shown in Fig. 8.3 is considered. The ultimate yield shear stress distribution is also shown in Fig. 8.3. It is clear that a shear band exists in the thinner portion with thickness t_1. The thicker portion has a distribution similar to Fig. 8.2b and contains the 'continuation' of the shear band of thickness t_1 over part of its thickness. The ultimate torque can be found as the sum of the torques contributed by the closed and open circuits.

$$T_u = T_u(\text{of the ring}) + T_u(\text{outside the half ring})$$

$$= \tau_y(2\pi a t_1)a + \tau_y\left[\pi a\left(\frac{t_2 - t_1}{2}\right)\right]\left[t_1 + \frac{t_2 - t_1}{2}\right] (2)$$

Fig. 8.2 a, b Yield stress distribution in cellular section

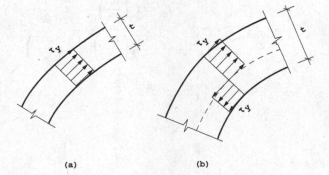

(a)   (b)

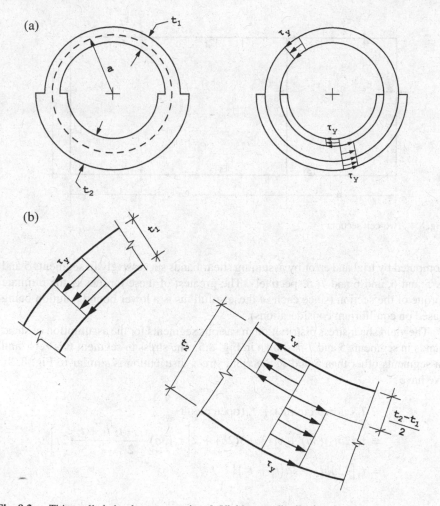

Fig. 8.3 a Thin-walled circular cross section. **b** Yield stress distribution

$$= \tau_y(2\pi a t_1)a + \tau_y\left[\pi a\left(\frac{t_2 - t_1}{2}\right)\right]\left[\frac{t_2 + t_1}{2}\right](2)$$

$$= \tau_y\left[2\pi a^2 t_1 + \frac{\pi a}{2}(t_2^2 - t_1^2)\right] \tag{8.2}$$

It may be noted that in the foregoing, the factor 2 in the second term accounts for the fact that the shearing stresses in the two orthogonal directions contribute equally to the torque in an open circuit. This fact is already explained in the derivation of Eq. (8.1).

As a second example, the two-cell section shown in Fig. 8.4 is considered. We assume that $t_1 > t_2$ where t_1 is the thickness of the horizontal segments and t_2 is the thickness of the vertical segments. The ultimate torque of the section can be

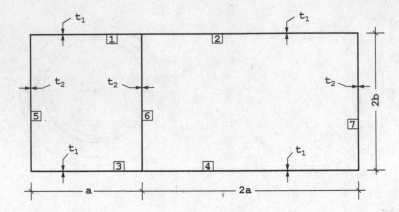

Fig. 8.4 Two-cell section

computed by trial and error by assuming shear bands successively in segments 5 and 7, 5 and 6, and 6 and 7, respectively. The greatest of these torques is the ultimate torque of the section (since each of these solutions is a lower bound solution being based on equilibrium considerations).

The yield shear stress distribution in various segments for the assumption of shear bands in segments 5 and 7 is shown in Fig. 8.5. The stress in segment 6 is zero, and in segments other than 5 and 7, the yield stress distribution is similar to Fig. 8.2b. We have

$$T_u = T_u(\text{closed circuit}) + T_u(\text{open cicuit})$$

$$= \tau_y(2bt_2)(3a) + \tau_y(3at_2)(2b) + 2\left[\tau_y(3a)\frac{t_1 - t_2}{2}\frac{t_1 + t_2}{2}(2)\right]$$

$$= \tau_y\left[12abt_2 + 3a(t_1^2 - t_2^2)\right]$$

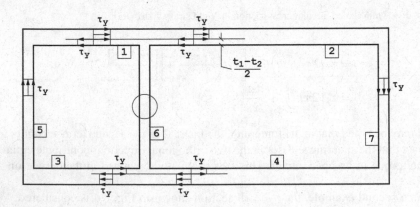

Fig. 8.5 Yield stress distribution with shear bands in 5 and 7

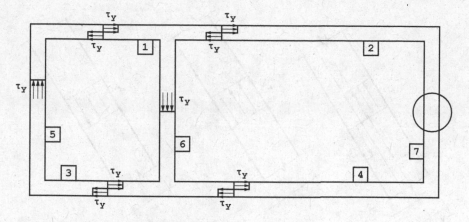

Fig. 8.6 Yield shear stress distribution with shear bands in 5 and 6

The yield shear stress distribution for the assumption of shear bands in segments 5 and 6 is shown in Fig. 8.6. We have

$$T_u = T_u(\text{closed circuit}) + T_u(\text{open cicuit})$$

$$= \tau_y[(at_2)(2b) + (2bt_2)(a)] + 2\left[\tau_y(3a)\frac{t_1 - t_2}{2}\frac{t_1 + t_2}{2}(2)\right]$$

$$= \tau_y\left[4abt_2 + 3a\left(t_1^2 - t_2^2\right)\right]$$

Finally, for the assumption of shear bands in segments 6 and 7, we have

$$T_u = \tau_y\left[8abt_2 + 3a\left(t_1^2 - t_2^2\right)\right]$$

Clearly, the greatest of these torques is

$$T_u = \tau_y\left[12abt_2 + 3a\left(t_1^2 - t_2^2\right)\right] \tag{8.3}$$

And, hence, the ultimate torque is associated with the yield shear stress distribution shown in Fig. 8.6.

8.4   Upper-Bound Solution

We consider only St. Venant torsion, and hence, warping must be free to occur everywhere. It has been seen that the neutral axis in a plastified segment can be located at a distance n_0 from the middle surface. Possible values of n_0 are shown in Fig. 8.7.

The warping in plastic torsion can be computed from the relationship

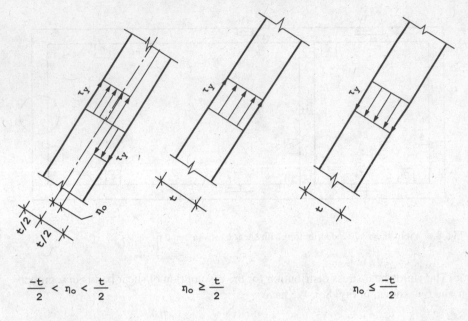

$$\frac{-t}{2} < \eta_0 < \frac{t}{2} \qquad\qquad \eta_0 \geq \frac{t}{2} \qquad\qquad \eta_0 \leq \frac{-t}{2}$$

Fig. 8.7 $n0$ values

$$\omega = \int \left[h_p + \frac{2n_0}{\left(1 - \frac{n_0}{R}\right)} \right] ds \tag{8.4}$$

where h_p is the tangential distance of an element ds from the pole and R is the radius of curvature of a curved contour. The continuity condition of the plastic (warping) mechanism gives

$$\oint \frac{2n_0}{\left(1 - \frac{n_0}{R}\right)} ds = -2A_i \tag{8.5}$$

for each cell i.

The upper-bound analysis is carried out in the usual manner. A collapse warping mechanism is to be found. The cross section is divided into redundant and determinate segments. The redundant segments are those that contain the shear bands (for which n_0 is to be calculated using the continuity conditions). The value of n_0 in the determinate segments must lie within the segments and can be obtained from the continuity condition at the junction with the redundant branches. Having found the n_0 values in all the branches, the ultimate torque can be calculated in the usual manner. This torque is the upper-bound value since it is based on kinematic (mechanism) considerations. If this value coincides with the highest lower bound, then the true ultimate torque has been found. If a plastic St. Venant shear flow, N, is defined, then

$$T_u = \tau_y \left[2 \int N r ds + \frac{1}{2} \int (t^2 - 4N^2) ds \right] \qquad (8.6)$$

As an application, we consider the section shown in Fig. 8.3. Using the continuity condition (Eq. 8.5), we have .

$$\frac{2n_0(\pi a)}{\left(1 - \frac{n_0}{a}\right)} + \frac{2\left(-\frac{t_1}{2}\right)(\pi a)}{1 + \frac{t_1}{2a}} = -2(\pi a^2)$$

where in the foregoing the value of n_0 in the determinate branch is assumed as $-\frac{t_1}{2}$. The foregoing equation gives the value of n_0 in the redundant branch.

$$n_0 = -\frac{2a^2}{t_1}$$

Now, $N_1 = N_2 = \frac{t_1}{2}$. Also, for continuity at the junction,

$$-\frac{2a^2}{t_1} < -\frac{t_1}{2}$$

Or

$$a \geq \frac{t_1}{2}$$

which is satisfied for a thin-walled section. Hence, the true mechanism has been found. The ultimate torque is given by Eq. (8.6).

$$T_u = \tau_y \left[2 \int_0^{2\pi} \left(\frac{t_1}{2}\right) a^2 d\theta + \frac{1}{2} \int_0^{\pi} (t_2^2 - t_1^2) a d\theta \right]$$

$$= \tau_y \left[2\pi a^2 t_1 + \frac{\pi a}{2} (t_2^2 - t_1^2) \right]$$

For a second example, the two-cell section shown in Fig. 8.4 is again considered. We have for the first trial.

$$n_{06} = 0 \quad n_{01} = n_{02} = n_{03} = n_{04} = -\frac{t_2}{2}$$

The continuity condition for the two cells gives

$$\left[2\left(-\frac{t_2}{2}\right) a + 2n_{05}(2b) + 2\left(-\frac{t_2}{2}\right) a \right] = -2(2ab)$$

$$\left[2\left(-\frac{t_2}{2}\right)(2a) + 2n_{07}(2b) + 2\left(-\frac{t_2}{2}\right)(2a)\right] = -2(4ab)$$

Hence,

$$n_{05} = -a\left(1 - \frac{t_2}{2b}\right)$$

$$n_{07} = -2a\left(1 - \frac{t_2}{2b}\right)$$

Also, for continuity at the junction,

$$-a\left(1 - \frac{t_2}{2b}\right) \le -\frac{t_2}{2}$$

$$a \ge \frac{t_2}{2\left(1 - \frac{t_2}{2b}\right)}$$

which is satisfied. Hence, the true mechanism has been found. The ultimate torque is given by, with the understanding that

$$N_5 = N_7 = \frac{t_2}{2}$$

$$T_u = \tau_y\left[2\left(\frac{t_2}{2}\right)(12ab) + \frac{1}{2}(t_1^2 - t_2^2)(6a)\right]$$
$$= \tau_y\left[12abt_2 + 3a(t_1^2 - t_2^2)\right]$$

The warping of the contour is given by

$$\omega_6 = 0$$

$$\omega_4 = (b - t_2)s_4$$

$$\omega_7 = -\frac{2a}{b}(b - t_2)s_7$$

$$\omega_2 = (b - t_2)s_2$$

$$\omega_1 = (b - t_2)s_1$$

$$\omega_5 = -\frac{a}{b}(b - t_2)s_5$$

$$\omega_3 = (b - t_2)s_3$$

8.5 Mixed Torsion

In actual situations, mixed torsion will govern especially near the points of restraint. Plastic analysis under mixed torsion is involved. A slight reduction in collapse load will result when plasticity under mixed torsion is accounted for.

Example 8.1 The cross section of a torsion member is shown in Fig. 8.8. Determine

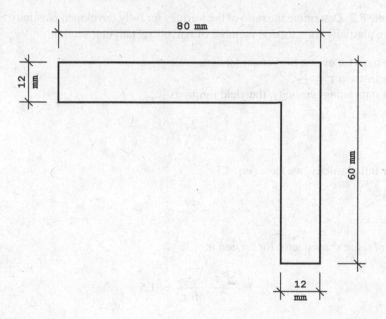

Fig. 8.8 Example 8.1

the ultimate torque corresponding to a fully plastic section. Yield stress in shear $\tau_y = 250\,\text{MPa}$.

The yield shear stress distribution is shown in Fig. 8.9. The distribution is divided into convenient shapes as shown. We have

Shape	Area (mm^2)	(Area) (N)	Lever arm from S (m)	Torque (Nm)
1. Triangle	$\frac{1}{2}(12)(6)$	9000	0.072	648.00
2. Trapezium	$\frac{1}{2}(6)(148)$	111,000	0.00308	341.88

(continued)

(continued)

Shape	Area (mm^2)	(Area) (N)	Lever arm from S (m)	Torque (Nm)
3. Parallelogram	(68)[6]	102,000	0.003	306.00
4. Trapezium	$\frac{1}{2}$(6)(108)	81,000	0.00311	251.91
5. Triangle	$\frac{1}{2}$(12)(6)	9000	0.052	468.00
6. Parallelogram	48(6)	72,000	0.003	216.00
				2231.79

The ultimate torque is thus

$$T_u = 2232 \, \text{Nm}$$

Example 8.2 Determine the ratio of the torques for fully developed plasticity to just starting plasticity for a torsion member of narrow rectangular section.

We have for elastic St. Venant torsion.
Shear stress $\tau = \frac{3T}{bt^2}$.
For impending plasticity, the yield torque is [2]

$$T_y = \frac{bt^2}{3}\tau_y$$

For full plasticity, we have (Eq. 8.1)

$$T_u = \frac{bt^2}{2}\tau_y$$

Hence, the shape factor for torsion is

$$= \frac{bt^2\tau_y}{2}\frac{3}{bt^2\tau_y} = 1.5$$

Example 8.3 Determine the shape factor in torsion for the two-cell section shown in Fig. 8.10. Yield stress in shear = 250 MPa.

The elastic St. Venant torsional analysis gives

$$\frac{95}{3}\psi_1 - 12.5\psi_2 = 8000$$

$$-12.5\psi_1 + \frac{115}{3}\psi_2 = 16,000$$

Giving $\psi_1 = 479.2$ and $\psi_2 = 573.59$.
Also,

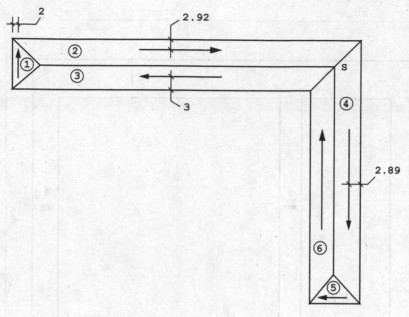

Fig. 8.9 Yield shear stress distribution

$$J = 2[479.2(4000) + 573.59(8000)]$$
$$= 13{,}011{,}040\,\text{mm}^4$$

The yield torque is thus

$$T_y = \frac{(250)(8)(13{,}011{,}040)10^{-3}}{573.59}$$
$$= 45{,}367\,\text{Nm}$$

The ultimate torque is given by Eq. (8.3)

$$T_u = 250[12(40)(50)(8) + 3(40)(20)(4)]10^{-3} = 50{,}400\,\text{Nm}$$

Shape factor for plastic torsion

$$= \frac{50{,}400}{45{,}367} = 1.11$$

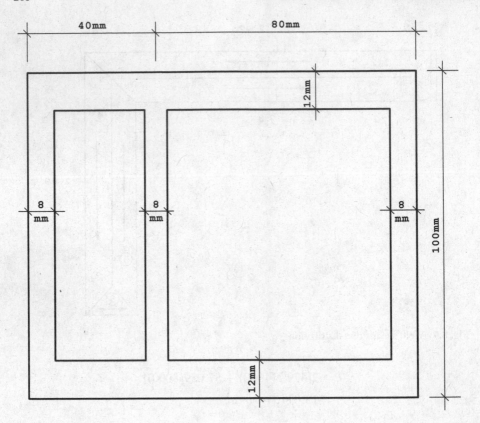

Fig. 8.10 Example 8.3

References

1. Gjelsvik, A.: The Theory of Thin walled Bars. Wiley, New York (1981)
2. Horney, M. R.: Plastic Theory of Structures. Thomas Nelson and Sons, Ltd, Nashville (1971)

Appendix
Analysis for Transverse Shear

Introduction

Efficient flexural analysis of thin-walled sections consisting of plate segments involves the determination of the shear center [1]. A brief account of the computation details has been presented in the following. Two examples of the calculation of the flexural center are also given to illustrate the details.

Example A.1 Determine the shear center of the twin-celled section shown in Fig. A.1. All plates are 4 mm thick. The center of gravity is C. $I_x = 448000$ mm^4.

Solution: Since x and y are the principal axis, shear flow calculations are to be carried out by applying unit shear forces parallel to these axes. However, since x-axis is an axis of symmetry, the shear center is located on this axis. Hence, a unit shear force parallel to the y-axis is all that is needed. Assuming two cuts as shown in Fig. A.2. and compressive forces at top and tensile forces at bottom we have, for a unit vertical shear force of 1 N, the shear stress is

$$\tau_1 = \frac{1}{(448000)(4)}[320 \times 20] = 0.00357 \text{ N/mm}^2$$

At the mid-point of plate 1–6, we have

$$\tau_{1-6} = \frac{(320)(20) + (80)(10)}{(448000)(4)} = 0.00402 \text{ N/mm}^2$$

The shear force in plate 1–6 using Simpson's rule.

$$\int \frac{q_0 ds}{t} = \frac{20}{3}[0.00357 + 4(0.00402) + 0.00357] = 0.1548 \text{ N/unit thickness}$$

These determinate shear forces are sketched in Fig. A.3 for all the plates.

K. Rajagopalan, *Torsion of Thin Walled Structures*, https://doi.org/10.1007/978-981-16-7458-7

Redundant shear flows q_1 and q_2 are assumed in the cells in the anticlockwise direction. We have

$$-\int\limits_{cell\ 1} \frac{q_0 ds}{t} = -0.1428 - 0.1548 - 0.1428 + 0.0833$$

$$= -0.3571$$

$$-\int\limits_{cell2} \frac{q_0 ds}{t} = -0.0358 - 0.0833 - 0.0358 + 0.0117 = -0.1432$$

For cell 1,

$$\frac{ds}{t} = \frac{80}{4} + \frac{40}{4} + \frac{40}{4} + \frac{80}{4} = 60$$

For cell 2,

$$\frac{ds}{t} = 4 \times \frac{40}{4} = 40$$

Thus,

$$\begin{bmatrix} 60 & -10 \\ -10 & 40 \end{bmatrix} \begin{Bmatrix} q1 \\ q2 \end{Bmatrix} = \begin{Bmatrix} -0.3571 \\ -0.1432 \end{Bmatrix}$$

Solving

$$q_1 = -0.006833 \text{ N/mm}$$

$$q_2 = -0.00529 \text{ N/mm}$$

The total forces in Newton in various plates are shown in Fig. A.4. The first sketch is obtained by multiplying $\int \tau_0 ds$ values by the thickness. The second is obtained by multiplying q_i by the respective lengths. The third sketch is obtained by adding the two. The position of the shear center from the left end is

$$e = (0.3932)(80) + (0.2588)(120) + (0.0416)(40) = 64.18 \text{ mm}$$

Example A.2 Determine the position of the shear center of the three-celled section shown in Fig. 6.18.

Three cuts are made as shown in Fig. A.5. We have $I_x = 2.7698 \times 10^{14} \text{ mm}^4$. For a unit vertical shear force of 1 N, we have

Plate 9–8

$$\tau_8 = \frac{(16290)(19)(14500)}{(2.7698 \times 10^{14})(19)} = 0.853 \times 10^{-6} \text{ N/mm}^2$$

Plate 8–7

$$\tau_8 = 0.772 \times 10^{-6} \text{ N/mm}^2$$

$$\tau_{8-7} = \frac{(16290)(19)(14500) + (1665)(21)(13667.5)}{(2.7698 \times 10^{14})(21)}$$
$$= 0.854 \times 10^{-6} \text{ N/mm}^2$$

$$\tau_7 = \frac{(16290)(19)(14500) + (3330)(21)(12835)}{(2.7698 \times 10^{14})(21)}$$
$$= 0.926 \times 10^{-6} \text{ N/mm}^2$$

Plate 9–10

$$\tau_{9-10} = \frac{(1665)(35)(13667.5)}{(2.7698 \times 10^{14})(35)} = 0.082 \times 10^{-6} \text{ N/mm}^2$$

$$\tau_{10} = \frac{(3330)(35)(12835)}{(2.7698 \times 10^{14})(35)} = 0.154 \times 10^{-6} \text{ N/mm}^2$$

Plate 10–6

$$\tau_{10} = \frac{0.154 \times 35}{12} \times 10^{-6} = 0.449 \times 10^{-6} \text{ N/mm}^2$$

$$\tau_6 = \frac{(3330)(35)(12835) + (14400)(12)(11170)}{(2.7698 \times 10^{14})(12)} = 1.031 \times 10^{-6} \text{ N/mm}^2$$

Plate 6–7

$$\tau_6 = \frac{(1.031)(12)}{(16)} \times 10^{-6} = 0.773 \times 10^{-6} \text{ N/mm}^2$$

$$\tau_7 = \frac{(3330)(35)(12835) + (14400)(12)(11170) + (1890)(16)(11170)}{(2.7698 \times 10^{14})(16)}$$
$$= 0.849 \times 10^{-6} \text{ N/mm}^2$$

Plate 7–5

$$\tau_7 = 0.926 \times 10^{-6} + \frac{(0.849)(16)}{21} \times 10^{-6} = 1.573 \times 10^{-6} \text{ N/mm}^2$$

$$\tau_{7-5} = 1.573 \times 10^{-6} + \frac{(11170)(21)(5585)}{(2.7698 \times 10^{14})(21)} = 1.798 \times 10^{-6} \text{ N/mm}^2$$

$$\tau_5 = 1.573 \times 10^{-6} \text{ N/mm}^2$$

Plate 6–4

$$\tau_{6-4} = \frac{(11170)(16)(5585)}{(2.7698 \times 10^{14})(16)} = 0.225 \times 10^{-6} \text{ N/mm}^2$$

Plate 1–4

$$\tau_4 = \frac{(14400)(12)(11170)}{(2.7698 \times 10^{14})(12)} = 0.581 \times 10^{-6} \text{ N/mm}^2$$

Plate 4–5

$$\tau_4 = \frac{(0.581)(12)}{16} \times 10^{-6} = 0.436 \times 10^{-6} \text{ N/mm}^2$$

$$\tau_5 = 0.436 \times 10^{-6} + \frac{(1890)(16)(11170)}{(2.7698 \times 10^{14})(16)} = 0.512 \times 10^{-6} \text{ N/mm}^2$$

Plate 5–3

$$\tau_5 = 1.573 \times 10^{-6} - \frac{0.512 \times 10^{-6} \times 16}{21}$$
$$= 1.183 \times 10^{-6} \text{ N/mm}^2$$

$$\tau_{5-3} = 1.183 \times 10^{-6} \frac{(1665)(21)(12002.5)}{(2.7698 \times 10^{14})(21)} = 1.111 \times 10^{-6} \text{ N/mm}^2$$

$$\tau_3 = 1.183 \times 10^{-6} - \frac{(3330)(21)(12835)}{(2.7698 \times 10^{14})(21)} = 1.029 \times 10^{-6} \text{ N/mm}^2$$

Plate 1–2

$$\tau_{1-2} = \frac{(1665)(35)(12002.5)}{(2.7698 \times 10^{14})(35)} = 0.072 \times 10^{-6} \text{ N/mm}^2$$

$$\tau_2 = \frac{(3330)(35)(12835)}{(2.7698 \times 10^{14})(35)} = 0.154 \times 10^{-6} \ \text{N/mm}^2$$

Plate 2–3

$$\tau_2 = \frac{(0.154)(35)}{19} \times 10^{-6} = 0.284 \times 10^{-6} \ \text{N/mm}^2$$

$$\tau_3 = 0.284 \times 10^{-6} + \frac{(16290)(19)(14500)}{(2.7698 \times 10^{14})(19)} = 1.137 \times 10^{-6} \ \text{N/mm}^2$$

The determinate shear stresses τ_0 for a shear force of 1 MN are shown in Fig. A.6. The values are in N/mm^2.

The integral $\tau_0 ds$ in N/mm for various segments of the open section are shown in Fig. A.7. We have for each of the three cells,

$$\frac{q_0 ds}{t} = -6947.69 - 2838.27 + 1532.79 + 10656 + 267.51 = 2670.34 \ \text{N/mm}$$

$$\frac{q_0 ds}{t} = -1532.79 - 38491.82 - 895.86 + 3351 = -37569.47 \ \text{N/mm}$$

$$\frac{q_0 ds}{t} = -3694.08 - 11574.05 - 245.31 + 4183.2 + 895.86 = -10434.38 \ \text{N/mm}$$

Thus,

$$2429.21 q_1 - 118.13 q_2 = 2670.34$$

$$-118.13 q_1 + 2696.31 q_2 - 118.13 q_3 = -37569.47$$

$$-118.13 q_2 + 2429.21 q_3 = -10434.38$$

Solving

$$q_1 = 0.4124 \ \text{N/mm}$$
$$q_2 = -14.1339 \ \text{N/mm}$$
$$q_3 = -4.9823 \ \text{N/mm}$$

The total forces in mega Newtons in various plates are shown in Fig. A.8. The shear center is located as

$$e = 0.370(1890) + (0.016)(16290) + (0.005)(11170)$$
$$- (0.278)(14500) - 0.244(11170) = -5740.69 \ \text{mm}$$

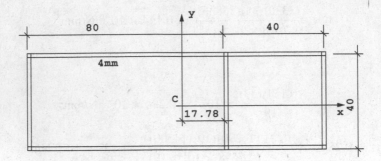

Fig. A.1 Twin cell section for Example A.1

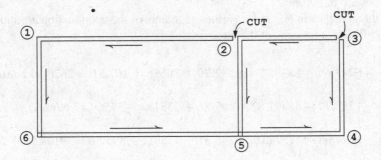

Fig. A.2 Schematic of shear flows in the cut section

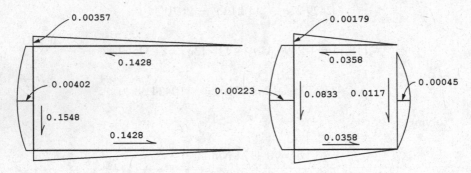

Fig. A.3 Shear forces per unit thickness in the cut section

It may be noted that this is the value of the shear center location given in Review problem 7 of Chap. 6.

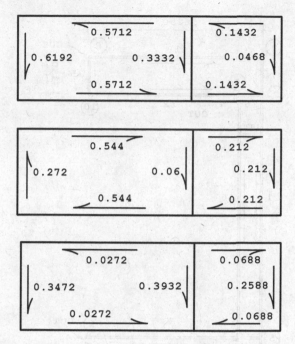

Fig. A.4 Shear forces in the cut section (top), corrective shear forces (middle) and actual shear forces (bottom)

Review Questions

1. Do the centers of shear and twist coincide?
2. A channel section with sheets and stringers are shown in Fig. A.9. Determine the shear center of this section.

Answers to Review Questions

1. When the shear center is defined via energy uncoupling between shear and torsion problems, the shear center is identical with the center of twist and is independent of the Poisson's ratio. If we consider pure flexure (no torsion) and pure (non-uniform in general) torsion (no flexure) as two systems, it can be shown using Betti's theorem that the center of twist coincides with the shear center [2].

    When the effect of warping shear is taken into account, things can be different.

2. The idealized structure of this example is extensively used in aerospace engineering wherein the booms (stringers) support the direct stress due to flexure and the plates support only the shear stresses arising from bending moment gradients. With this idealization, the moment of inertia about the centroidal axis

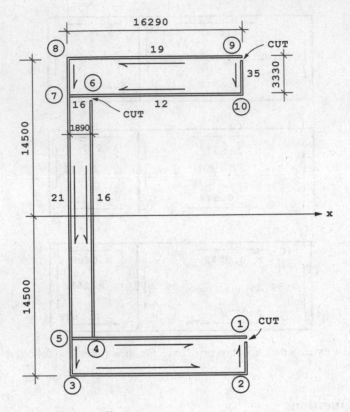

Fig. A.5 A section with three cells

of symmetry is

$$= 2(120)(60)^2 + 120(20)^2 + 60(60)^2$$
$$= 1392000 \text{ mm}^4$$

The shear stress at 1, for a vertical shear force of 1 N, is

$$= \frac{1}{1392000}(60)(60) = 0.0025862 \text{ N/mm}$$

And, the shear force is in the plate

$$= (0.0025862)(70)$$
$$= 0.181 \text{ N}$$

The shear stress at 2 is

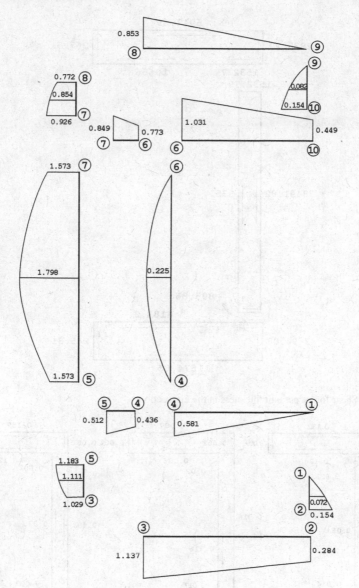

Fig. A.6 Shear stresses in the cut section

$$= \frac{1}{1392000}[(60)(60) + (120)(60)]$$
$$= 0.0077586 \text{ N/mm}$$

This is constant (though the plate is a web, i.e., parallel to the load) because of the sheet-stringer assumption. The shear force over this plate segment is

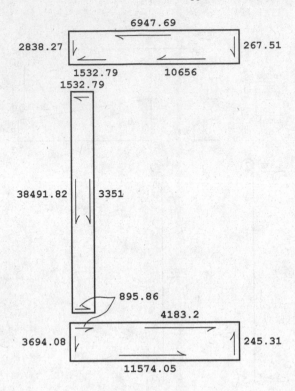

Fig. A.7 Shear forces per unit thickness in the cut section

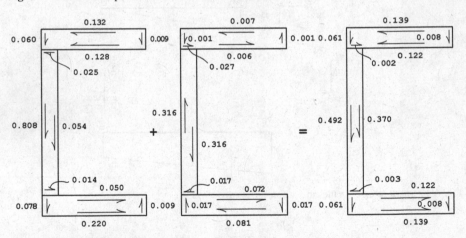

Fig. A.8 Shear forces in the cut section (extreme left), corrective shear forces (middle) and actual shear forces (extreme right)

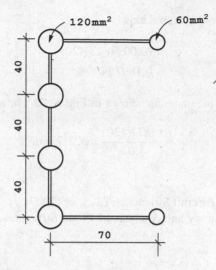

Fig. A.9 A channel section with sheets and stringers for review question A.2

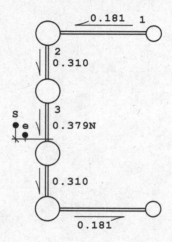

Fig. A.10 Shear forces on the sheets

$$= (0.0077586)(40)$$
$$= 0.310 \, \text{N}$$

The shear stress at 3 is

$$= \frac{1}{1392000}[(60)(60) + (120)(60) + (120)(20)]$$
$$= 0.0094827 \, \text{N/mm}$$

The shear force up to the neutral axis is

$$. = (0.0094827)(20)$$
$$= 0.1897 \text{ N}$$

The shear forces on the plates are shown in Fig. A.10. The shear center is

$$e = \frac{(0.181)(120)}{1.0} = 21.7 \text{ mm}$$

References

1. Megson, T.H.G.: Aircraft Structures. Elsevier (2007)
2. Rivello, R.M.: Theory and Analysis of Flight Structures. McGraw Hill, New York (1969)

Index

Printed in the United States
by Baker & Taylor Publisher Services